THE SCIENCE OF
STAR TREK

THE SCIENCE OF

STAR TREK

THE SCIENTIFIC FACTS BEHIND THE VOYAGES IN SPACE AND TIME

MARK BRAKE

AUTHOR OF *THE SCIENCE OF STAR WARS*

Skyhorse Publishing

Skyhorse Publishing books may be purchased in bulk at special discounts for sales promotion, corporate gifts, fund-raising, or educational purposes. Special editions can also be created to specifications. For details, contact the Special Sales Department, Skyhorse Publishing, 307 West 36th Street, 11th Floor, New York, NY 10018 or info@skyhorsepublishing.com.

Skyhorse® and Skyhorse Publishing® are registered trademarks of Skyhorse Publishing, Inc.®, a Delaware corporation.

Visit our website at www.skyhorsepublishing.com.

10 9 8 7 6 5 4 3 2 1

Library of Congress Cataloging-in-Publication Data is available on file.

Cover design by David Ter-Avanesyan
Cover image by Shutterstock

Print ISBN: 978-1-5107-5788-2
Ebook ISBN: 978-1-5107-5789-9

Printed in the United States of America

To my grandson, Nate, who dreamt of starships at just a day old.

CONTENTS

THE SCIENCE OF
STAR TREK

INTRODUCTION

"If you want your children to be intelligent, read them fairy tales. If you want them to be very intelligent, read them more fairy tales."

—Albert Einstein

"I don't think there is any world where *Star Trek* is anything but a progressive, liberal vision of the future in which big government is a good thing, and we can all get along. It's a utopian ethos that is a result of one world government, and not exceptionalism of any particular country."

—Mark A. Altman, *The Washington Post* (2016)

"*Star Trek* was an attempt to say that humanity will reach maturity and wisdom on the day that it begins not just to tolerate, but take a special delight in differences in ideas and difference in life forms. If we cannot learn to actually enjoy those small differences, to take a positive delight in those small differences between our own kind, here on this planet, then we do not deserve to go out into space and meet the diversity that is almost certainly out there."

—Gene Roddenberry, creator of *Star Trek: The Original Series*

GOSPEL FOR OUTER SPACE

It's an honor to be alive, don't you think? To witness the grace, beauty, and complexity of the cosmos. A countryman of mine, the great but sometimes melodramatic actor Richard Burton, once said he wanted to end it all because of the beauty. Contemplating the sublime was simply too much for him. In contrast, *Star Trek* rejoices in that beauty. The Kirk-spoken title sequence of each episode in *The Original Series* suggests as much: "Space: the final frontier. These are the voyages of the Starship Enterprise. Its five-year mission: to explore strange new worlds. To seek out new life and new civilizations. To boldly go where no man has gone before!" (The introduction was, of course, changed for *Star Trek: The Next Generation* to the more inclusive "where no one has gone before" and "its continuing mission," to reflect ongoing missions.)

First conceived by Gene Roddenberry as early as 1964, *Star Trek* has been a cult sensation ever since. Today, the franchise encompasses a broad range of offshoots including novels, comics, figurines, games, and toys. Museum exhibits of the props from the franchise travel the globe. For the decade between 1998 and 2008, an attraction based on *Star Trek* played in Las Vegas. The constructed language Klingon, spoken by the fictional Klingon race in the Star Trek Universe, was created for the franchise and is actually spoken by some devotees.

The franchise has generated between $10 to $20 billion in revenue, making *Star Trek* one of the highest-grossing media franchises in history. Not only is *Star Trek* recognized for its influence on the science and culture beyond science fiction, but it's also known for its progressive civil rights stances. *The Original Series* included one of television's first multiracial casts. From everyday science and tech to the quest to travel among the stars, *Star Trek* has impacted the way we think, the way we live, and the way we use tech on a daily basis.

Lawrence Krauss's wonderful book *The Physics of Star Trek* was written way back in 1995. It's reasonably fair to say that Lawrence's book kick-started the subgenre of "science of" books which are now a popular feature of publishers' catalogs. And here I am, writing my twenty-third "science of" book, which is partly about the twenty-third century, among others. Like Lawrence, I am lucky enough to be in the generation that first witnessed the appearance of *Star Trek* on our television screens in the 1960s. I must admit that I've never attended a Comic-Con decked out as a member of the Borg, but I have nonetheless had a long and wide experience of the franchise over the last six decades or so.

In that time, *Star Trek* has not only helped us imagine the science and cultures of future human societies, but it has also helped us imagine what those societies themselves might look like. So, in this book, you will find chapters not only on the related science and tech of *Star Trek*, but also the more social science side of the franchise. *Star Trek* is, of course, science fiction. And all science fiction can be thought of as being about the cultural shock of discovering our marginal position in an alien Universe, as revealed by the advance of science. *Star Trek* is an attempt to put the stamp of humanity back on to the Universe. To boldly go.

Furthermore, science fiction, like *Star Trek*, is concerned with the relationship between the human and the nonhuman. On the surface, *Star Trek* seems to have a bewildering number of themes: aliens and time machines, spaceships and cyborgs, utopias and dystopias, androids and alternate histories. But, on a more thoughtful level, we can identify four main themes: space, time, machine, and monster. Each of these themes is a way of exploring the relationship between the human and the nonhuman. Taking a closer look at these themes will enable a clearer understanding of the ways in which *Star Trek* works, and what the franchise has to say about science and society.

SPACE

The space theme sees the nonhuman as some aspect of the natural world, such as vast interstellar spaces in which the Federation travels, or the alien, which can be seen as an animated version of nature. Here we look at topics such as space travel, the science of exoplanets, and why we even call them space "ships" in the first place.

TIME

The time theme portrays a flux in the human condition brought about by processes revealed in time. Tales on time often focus on the dialectic of natural history, so they are of particular relevance to evolution and biology. In this part, we look at the ways *Star Trek* deals with topics such as alternate histories, how to leave footprints in time, and the history of *Star Trek* itself.

MACHINE

The machine theme deals with the "man versus machine" motif, including robots, computers, and artificial intelligences. Dystopian tales are part of the "man versus machine" theme; it is the social machine in which the human confronts the nonhuman in such cases. This part has entries on *Star Trek*'s machine motif, including machine slavery, the Dyson Sphere, and the replicator.

MONSTER

In monster tales there is often an agency of change, such as the cosmic evolution in "The Chase," which leads to the development of different humanoid aliens. We also look at the monstrous nature of war, in which humans become more bestial, and how the political use of tech may mean modern humans are becoming more Borg.

Naturally, many of the scientific wonders of the Star Trek Universe sit at odds with our current conceptual understanding of the cosmos. But the point of this book is not to take *Star Trek* tales scientifically literally. *Star Trek* is science fiction, after all. Nor will you find here an obsession with the mere inventions alone, a kind of commodity fetishism about the guns, gadgets, and transporters that litter the franchise. They are mere decoration. No, this book looks at the bigger picture. This book takes the long view, the larger-than-life scientific and cultural contexts, which act as world-shaking scenarios to the Star Trek Universe.

Live long and prosper.

PART I
SPACE

WHAT CAN *STAR TREK* TELL US ABOUT ALIENS IN OUR MILKY WAY?

"Almost certainly there is enough land in the sky to give every member of the human species, back to the first ape-man, his own private, world-sized heaven—or hell. How many of those potential heavens and hells are now inhabited, and by what manner of creatures, we have no way of guessing; the very nearest is a million times farther away than Mars or Venus, those still remote goals of the next generation. But the barriers of distance are crumbling; one day we shall meet our equals, or our masters, among the stars."

—Arthur C. Clarke, *2001: A Space Odyssey* (1968)

GALAXIES

"Space. The Final Frontier. The USS *Enterprise* embarks on a five-year mission to explore the Galaxy." Like island Universes, Galaxies are swarms of Suns, adrift in the ocean of space-time. Each a vast collection of stars, dust, gases, and matter, a Galaxy is bound by gravity. And, collectively, Galaxies are the cells that make up the large-scale structure of the cosmic body of the Universe. Natural events in galactic dynamics occupy tens of millions of years, and yet our modern notion of a Galaxy is barely one hundred years old.

On April 26, 1920, at the Smithsonian Museum of Natural History, astronomy's Great Debate took place. At stake was the human measure of the very scale of the Universe itself. During the Debate, one astronomer,

Harlow Shapley, argued that our Milky Way Galaxy was the entirety of the Universe. His opponent, Heber Curtis, contended that the great spiral nebula in Andromeda and other such "nebulae" were in fact separate Galaxies, or "island Universes," a term originally coined by eighteenth-century German philosopher Immanuel Kant, who was one of the very first to believe that the "spiral nebulae" were extragalactic. It wasn't until later in the 1920s that a perhaps more famous American astronomer, Edwin Hubble, showed that many nebulae, thought to be clouds of dust and gas or single stars in formation, were in fact Galaxies beyond our Milky Way.

For many years, the number of Galaxies in the observable Universe was thought to be around two hundred billion, but in October 2016, *The New York Times* reported that astronomers now estimate the cosmic Galaxy number at two trillion. At the very least. *Star Trek: Picard* fact checks this new Galaxy number in episode four "Absolute Candor" when Doctor Agnes Jurati muses over the nature of space: "There are a septillion known planets, so maybe [space] should be called "vast quantities of stuff."

The sheer scale of the cosmos is terrifying for some. It certainly worried Victorian poet Alfred Tennyson, even though little was then known about modern cosmology. In 1855, Tennyson wrote his poem "Vastness" in which he declared, "This poor Earth's pale history runs—what is it all but a trouble of ants in the gleam of a million million of Suns?" The population of the Milky Way doesn't quite reach the heady heights of Tennyson's million million Suns. The old British billion of a million million has since lost its battle with the American billion of a mere thousand million. The Milky Way, in which our Sun sits, is thought to contain between one hundred and four hundred billion Suns, and more than one hundred billion planets. Rather than being troubled like Tennyson by this vastness, *Star Trek*'s unremitting missionaries explore these strange new worlds, aiming to seek out new life and new civilizations.

STAR TREK: A CROWDED GALAXY

In the Star Trek Universe, the Milky Way is replete with alien life, of course. The Galaxy is divided into four quadrants: Alpha, Beta, Gamma, and Delta (the first four letters of the Greek alphabet). Each quadrant roughly comprises one quarter of the Star Trek Galaxy, and each quadrant is divided into thousands of sectors. As we well know, the four great powers in the Alpha quadrant are the United Federation of Planets, the Cardassian Union, the Klingon Empire, and the Romulan Star Empire, though admittedly these last two imperial powers hold most of their territory in the Beta quadrant. The galactic core is the intersection of all four quadrants. (Real astronomers also divide our Galaxy into quadrants. What's the main difference between the real and fictional quadrant systems? The galactic quadrants in astronomy are based around a perpendicular access that runs through the Sun, while the *Star Trek* system is far less geocentric; it's based around an axis that runs through the galactic center. In short, *Star Trek* is less parochial and more progressive!)

Watching *Star Trek* makes us wonder something quite profound. If our Milky Way was anything like the Star Trek Galaxy, might it be possible, if our Galaxy also teems with space-faring civilizations, to have isolated and unvisited worlds like ours? Could it be that Earth simply remains, as yet, undiscovered by alien intelligence? The encouraging scientific answer to this question is a resounding yes!

Here's a summary of what we are about to argue: scientific projections suggest that, if there are alien planet-hopping species in the Galaxy, they could spread across the Milky Way quite swiftly, cosmically speaking. Why have we found no undeniable evidence of extraterrestrial visits to Earth? Because galactic settlement happens in waves. And humanity has arisen on a relatively obscure planet during a local lull in galactic exploration.

SEA OF WATER, SEA OF SPACE

Before we look in more detail at the Galaxy's sea of space, consider the case for our Earthly seas of water, and the story of HMS *Bounty*. Launched in 1784, the Bounty was a small merchant vessel that the British Royal Navy bought for a botanical mission. Rather famously, the Bounty was sent to the Pacific Ocean under the command of Captain William Bligh, but the mission was never completed. A mutiny led by acting lieutenant Fletcher Christian seized control of the ship from Bligh, and the mutineers settled on local islands. Some readers may recall movie portrayals of the mutiny of the Bounty, where Christian is played by Marlon Brando (1962), and Mel Gibson (1984). In *Star Trek IV: The Voyage Home*, "HMS BOUNTY" is roughly painted in large red letters along the side of the Bird-of-Prey, with Kirk writing in his log, "like those mutineers of five hundred years ago, we too have a hard choice to make."

In January 1790, nine mutineers from the Bounty, along with eighteen Tahitians and a small child, set foot on Pitcairn Island, one of the most remote but habitable places on the planet. Adrift in the South Pacific Ocean, with hundreds of miles of sea to neighboring islands, Pitcairn is its own island Universe. The mutineer group were the first people Pitcairn had seen in some time. The island had not had human occupation since the fifteenth century, having been previously inhabited by Polynesians until the ecology became simply unsustainable. But here's the thing: after the arrival of the Bounty party in 1790, it was almost another generation before any other ship dropped anchor at Pitcairn Island. Though the Bounty settlers saw ships sailing past in the distance, no one else set foot on land.

It might come as a surprise to some readers to learn that New Zealand was one of the last major landmasses settled by humans. Various techniques including radiocarbon dating, deforestation evidence, and mitochondrial DNA nuance in local Māori populations suggest that New Zealand was first settled by Eastern Polynesians as late as 1300 AD. That's really something when you consider the fact that archeologists and population geneticists believe Indigenous Australians inhabited the Australian continent for perhaps as many as 65,000 years prior to European "discovery." (To give a measure of how close the two landmasses are, during the Australian bushfires of late 2019 and early 2020, the skies over New Zealand turned at times orange and other times dark and gloomy from the residual smoke.)

Such are the unusual dynamics of human occupation. Easter Island (Rapa Nui) is the most remote inhabited island. It took humans almost all of history to get there. The Polynesians were the supreme open ocean-goers in world history, and their skill at navigation using double-hulled canoes over vast expanses of sea is one of human history's greatest triumphs. They settled New Zealand, Hawaii, and many other islands in the remote Pacific.

The tales of Pitcairn and New Zealand are far from unique. Across the South Pacific are thousands of islands essentially lost in the millions of square miles of sea. Many of these islands may be mere rock and coral, of course. Others, like Pitcairn, may boast habitable ecologies but no human inhabitants. But, like the stars across interstellar space, these islands are constellations of potential settlement for those motivated enough to navigate the space between.

AND SO TO THE STARS

How does this example of Earth's South Pacific apply to the Star Trek Galaxy and our actual Milky Way? The parallels are striking. As we said earlier, the estimate of the number of stars in our Galaxy puts the stellar population at as many as four hundred billion Suns. And the latest estimates from NASA's exoplanet hunters suggest the Galaxy harbors more than ten billion habitable rocky worlds within the seas of the Milky Way. Like the islands that speckle Earth's oceans, these exoplanets are possible living way stations. They could provide a series of stepping-stones for any species with the outward urge to migrate across galactic space.

This comparison is pretty crucial to our argument. Europeans finally discovered that Polynesians had spread across thousands of miles of the southern Pacific on simple watercraft, happily sailing along at just a few knots. Similarly, galactic migration could be carried out with relatively simple spacecraft and, naturally, sufficient oodles of cosmic time. (When we look back at the Apollo missions from the relatively technologically sophisticated twenty-first century, we are often amazed by the crude and simple spacecrafts that first took astronauts out to the Moon.)

FERMI'S PARADOX

This simple state of affairs also provides a new answer to Fermi's so-called Paradox. As legend would have it, in response to the question of extraterrestrial intelligences, Nobel Prize–winning Italian physicist Enrico Fermi said over lunch with colleagues in 1950, "but where is everybody?!" Where indeed were all the spacefaring species that, according to most science fiction, were meant to be clogging up the cosmos? Fermi's point was this: unless alien life was rare, an argument that biologists sometimes make, tech-savvy extraterrestrials should have spread through the Galaxy. And yet we see no credible evidence (save the odd story of Greys creeping up behind stoned hippies at rock festivals). Fermi was known for his quick wit. He'd won a Nobel Prize, after all. And over lunch and in one quip he seemed to have figured out that migration into the Milky Way could be done in an instant (as long as that instant was a few million years)!

Ever since 1950, physicists have been refining their interpretations of Fermi's Paradox. Take, for example, the fact of the continuing absence of aliens on our planet. This absence leads some scientists to conclude that there are no other technological civilizations in our Galaxy. Nor indeed, they suggest boldly, have there ever been. It seems that the crucial factor in this glum scenario is the assumption that it would take a relatively short amount of time for an extraterrestrial species to migrate across the Milky Way's hundred thousand light-year span. Even if, as the argument goes, the alien spacecraft was propelled along at plodding sub-light speed.

According to associated calculations, in a few million years, those aliens possessed with an outward urge could have visited every last nook and cranny of the local galactic cosmos. As our solar system has been here for almost five billion years, and latest estimates put the age of the Milky Way at around ten billion years, there has been plenty of time for those tech-savvy extraterrestrial species to visit all habitable planets.

FREESTYLING FERMI

In most of these refinements to Fermi's lunchtime quip, there is a pretty constant scenario, and it goes something like this: stars and their planets become inhabited, assuming they weren't so already, then act as the next-stage stepping-stone for further onward migration to new systems. And yet the scenarios always seem to assume the same thing about the scope of the tech used for galactic travel, and that sentient beings can actually survive the journey. What do we know today about these scenarios for interstellar travel?

Well, for one thing, the scientific consensus is that even traveling at a modest 10 percent of light speed will need some pretty nifty tech such as fusion-bomb propulsion, gargantuan laser-driven light sails, and spacecraft design to protect ships from high-speed effects. Design features include shielding of the hull from the constant attrition of interstellar gas atom impacts and starship-smashing rock debris, which packs the punch of a bomb at any reasonable fraction of light speed. Naturally, slower speeds are safer, but come with interstellar journey times of centuries or millennia, and we have yet to learn how to keep a ship's crew alive for such time spans.

There are other takes on Fermi's Paradox, of course. A fascinating one is the idea that, unlike humans, alien civilizations have no interest in reaching other stars. No outward urge. No sense of curiosity, weirdly. None of them. You can see that, based on this premise, the whole mission of Milky Way migration melts away from the get-go. For *Star Trek* fans, who are well aware of the possibilities of the rich variety of life in our Galaxy and beyond, it's easy to see that this stay-at-home reading of Fermi is something of a monocultural mistake. Would all variations of extraterrestrial species really be so like-minded? Heavens, we can't even get the different branches of homo sapiens to agree on a number of matters. Besides, even if most aliens were against dipping their toes in galactic waters, all it would take is a single curious species to take the plunge and spread their seed across the light years of the Milky Way.

Indeed, the historical debate on Fermi is rich in alternate hypotheses. One solution is based on the simple matter of the cost of space travel. It may be that the expense of reaching the stars, and maintaining that interstellar reach across space, is simply too high, even for a technologically advanced species who have carefully planned their conquest of space from the start. Or it may be that population growth does not motivate a species to take to the stars. This last solution would be especially true perhaps for an extraterrestrial race that rationally plans ahead, developing a sustainable ecology on its home planet. Why travel when the other aliens' grass isn't necessarily greener, or whatever color alien grass might be, assuming they actually had grass, of course. (Thinking about this, would the, hopefully, imminent green revolution on Earth result in the cancellation of human space programs? The revolution might mean space exploration has no point other than scientific curiosity, which might prove tricky as many venture capitalists are merely looking for some kind of bang-for-their-buck bonanza.)

Then, there's the "great barrier," the notion that there are factors that limit alien conquests of space. Maybe it could be a failure to reach that green revolution, with rampant industrialization leading to an entropic and pollutive decay of the planet and all technological progress. Of course, there are plenty of natural barriers, too. It's hard to explore space when you're hampered by natural catastrophes such as supernovae or the terrible and creeping advance of a nearby black hole, which coldly eliminates some potential spacefarers from galactic life. Readers may recall *Star Trek: The Next Generation*'s episode "The Inner Light." It features a civilization on the non-federation planet Kataan, whose exploration of space was terminated by their Sun going nova, exterminating all life in their system. And yet one feels these extreme examples are the exception that proves the rule. (It would be remiss of me not to mention here what's sometimes called the zoo hypothesis. According to this rather fringe and somewhat science fictional theory, humans are being kept in the dark by alien powers. We remain isolated because that's the way ET wants it. It's an extension of the usual political terrestrial paranoia played out in the stars, rather than the corridors of power.)

BACK TO THE BOUNTY

Star Trek could be closer to the truth than "real" scientific hypotheses. Why? None of the above perspectives on Fermi's Paradox explain what we know to be true about the transient and irregular nature of human habitation on the South Pacific islands. Bearing in mind this invaluable lesson from planet Earth, for all we know there could be regions of the Galaxy, archipelagos of cosmic life, where interstellar contact is common. For both extraterrestrial and terrestrial examples, the parameters may differ in detail but have similar net effects, from the choices that influence where a ship pulls into port to the time taken before explorers feel the need to push farther across the seas of water or space. Like the case of the South Pacific, pulses of cosmic exploration, discovery, and settlement could ebb and flow across the Milky Way. The recent rise of homo sapiens could simply have happened during one of those cosmic downtimes, when galactic migration is in quiescence.

But might Earth have been visited in the past? In 2012, I wrote a book for Cambridge University Press called *Alien Life Imagined*, telling the tale of the 2,500-year history of the human portrayal of imagined alien life. Yes, that's how long we've been dreaming up extraterrestrials. The book takes examples from the history of science, philosophy, and science fiction. One of the stand-out quotes on the question of whether Earth may have been visited in the past comes from American anthropologist Loren Eisley in her 1957 book, *The Immense Journey*:

> So deep is the conviction that there must be life out there beyond the dark, one thinks that if they are more advanced than ourselves they may come across space at any moment, perhaps in our generation. Later, contemplating the infinity of time, one wonders if perchance their messages came long ago, hurtling into the swamp muck of the steaming coal forests, the bright projectile clambered over by hissing reptiles, and the delicate instruments running mindlessly down with no report . . . in the nature of life and in the principles of evolution we have had our answer. Of men elsewhere, and beyond, there will be none forever.

Eisley belongs to that group of evolutionists who are impressed by the incredible improbability of intelligent life ever to have evolved, even on Earth. Over the years since Eisley's book, a number of academics have considered the possibility of searching for archeological artifacts, evidence of an alien visitation to Earth and our solar system. Some readers may also remember the *Star Trek: The Next Generation* episode "Time's Arrow" in which the Enterprise is recalled to Earth to investigate an alien visit to San Francisco, five hundred years before, only to discover the severed head of Data! It's unlikely that archeologists will meet with this degree of difficulty, but who knows? The scope of such archeological searches are hard to define, while some academics also question whether we could actually tell if there had been an earlier industrial civilization on Earth.

This is no exaggeration. Most planetary scientists believe that time would easily eradicate any signs of past technological life on the planet. Given a stretch of one million years, the evidence would amount to very little, such as the odd barely detectable synthetic molecule, plastic, or trace radioactivity. Should our own civilization end tomorrow, the same would be true of any evidence left behind. Modern urbanization covers only 1 percent of the Earth's surface. A vanishingly small target for any paleontologists of the future. So, if an industrial civilization like ours had existed a few million years ago, we wouldn't know it.

FROM PITCAIRN TO PICARD

Let's use the South Pacific as a model for the way in which planet-hunting aliens may make their way across the Milky Way. One of the main things to keep in mind here is the actual motion of the Galaxy. When many of us think about making our way across space, we often visualize static positions for the stars, as that's how we see them when we stargaze from Earth. But the Galaxy is in constant motion. Remember that the Sun is in orbit around the center of the Milky Way. The galactic year, as it's called, is between 225 and 250 million terrestrial years, as our solar system is traveling at an average speed of 514,000 mph along its trajectory around the galactic center.

Imagine taking a selfie of the stars in the solar neighborhood. Let's say the selfie has a circular picture frame with a radius of two hundred light-years of the Sun. Those neighborhood stars are in motion. They behave essentially like the moving molecules of a gas. Relative to the fixed center of our picture frame, each star may be moving slowly or swiftly in any random direction through local space.

If we now zoom the selfie out farther (this is an incredibly long selfie stick we're using) we begin to see the large-scale pattern. At the scale of thousands of light-years, the grand sweep of the galactic year becomes evident, as the mass of stars wheel in a shared orbital motion around the Milky Way's core. Naturally, those stars much closer to the core take less time to complete their year. Much farther out are rapidly moving halo stars dipping in and out of the galactic disc, which form a spherical swarm around the Galaxy's plane.

MODELING GALACTIC MIGRATION

Let's put ourselves in the position of a future spacefaring human civilization. If we are searching the local solar neighborhood for potential stars to explore, our best bet closest target will change over time. For example, at the moment the closest star to our solar system is Proxima

Centauri. Proxima is an excellent target star, as the exoplanet Proxima Centauri b sits in the habitable zone in orbit around the red dwarf star Proxima. So not only is Proxima the closest star to the Sun, at only 4.24 light-years from Earth, it also has the closest exoplanet to Earth. Even better, in around 10,000 years, Proxima will be even more proximal! By that time, it will be only 3.5 light-years distant. That's quite a potential saving in both interstellar travel time and energy. Furthermore, were our spacefaring civilization to wait until 37,000 years into the future, they'd find their nearest neighbor would be Ross 248, a red dwarf star that would then be only 3 light-years distant.

How do astronomers simulate such a dynamic situation? For their stellar cartography, they model a 3d box of stars, whose movements mock a small part of the Milky Way, as we described above. The simulation then models a frontier settlement. It does this by marking a coherent series of stars as first hosts to planet-hopping civilizations. Thinking about Pitcairn, civilizations have finite life spans, so a star-planet system can become unoccupied in time too. Another keyed in factor for the model is that a spacefaring civilization will have a waiting period before it's able to launch a migration mission to its neighboring star. The model is flexible. All the above factors can be adjusted to further refine the simulation. But for a considerable range of values, a migration frontier nonetheless advances through galactic space. It's the speed of this advance that is crucial to validating certain solutions to Fermi's riddle.

Consider a worked example of the model. Imagine we use the 3d box of stars in the following manner. Our timescale will be 10 million years. Our theater of exploration will be 10,000 potentially settleable worlds within a cuboid 464 light-years long. In this model, inhabitable worlds outnumber habitable ones by a factor of over 20. Again, the galactic dynamics mimic the random motion of the molecules of a gas. But it's important to note that the probes and spacecraft from the cultures populating this space travel 100 times faster than the stars themselves. At the end of the 10 million years, 6,948 of the 10,000 systems had been visited by tech, and yet only 403 boasted active settlements. The remaining 3,052 systems, like Earth, were left unvisited. So far. What's more, the working example created 11 different "empires," which comprised 10 settled systems each.

The model's findings are fascinating. For one thing, the gas-like movement of the stars means that even the most sluggish spacecraft (say twice as fast as Voyager 1) would mean a migration frontier would cross the Milky Way within a billion years. If other dynamic parameters are tweaked, the frontier time span gets shorter. In short, Fermi was right. At least in this much: it's easy to fill the Galaxy with alien life. But the question remains, how full will the Galaxy become, and in what migrational pattern? The answer to this question depends not only on the number of habitable worlds in the Milky Way, but also on the lifetime of settlement civilizations enduring on a frontier world.

On the one hand, it's simple to make the Galaxy devoid of alien life. All one need do with the model is assume civilizations endure for around 100,000 years, and also greatly reduce the number of habitable planets (all the data from NASA and other agencies totally undermine this second assumption, of course). On the other hand, and given that same data from NASA, it's possible to adjust the model's parameters so that the local cosmos is replete with Milky Way migration and the resulting spacefaring settlements. Indeed, if NASA is right about the sheer number of habitable worlds, it matters little how long frontier settlements last. As long as they keep their space traveling tech, sufficient numbers of their civilizations would continue to planet hop and fill the Galaxy.

Consider now the more fascinating middle way. Sitting betwixt the two extremes mentioned above is perhaps the most interesting scenario, both for our actual Galaxy and for students of *Star Trek*. When the regularity of settlements of habitable planets is a middling number between the two extreme cases, the pattern of settlements creates clusters of systems. These clusters are like an archipelago of islands in the South Pacific. Some stars are continually visited or resettled, by pulse after pulse of galactic explorers. And yet these archipelagos are also surrounded by regions of unsettled space, galactic places that are simply a little beyond worth exploring. One is reminded here of Douglas Adams's famous words in *The Hitchhiker's Guide to the Galaxy* describing our Sun: "Far out in the uncharted backwaters of the unfashionable end of the western spiral arm of the Galaxy lies a small unregarded yellow Sun."

GALACTIC ARCHIPELAGO

Even more exciting for *Star Trek* and science fiction enthusiasts is the fact that this galactic archipelago model accounts for our current situation here on Earth. For instance, imagine we plug the following values into the model's parameters: One, we assume planetary civilizations endure for around a million years, and two, we assume that only 3 percent of star systems are actually habitable. When these values are used, the model returns a projection that a habitable world like ours has a 10 percent chance of not having been visited in the past million years or so. In short, it's hardly a surprise that we should find ourselves isolated and unvisited.

Excitingly, this model also suggests that in other parts of our Galaxy there are long strands of stellar archipelagos where extraterrestrial visitors and alien neighbors are perfectly possible. No one is saying our Galaxy houses Cardassians, Klingons, and Romulans, but neither do we need to render the model extreme to conjure a believable scenario for alien life to be commonplace. All we need is an eye for detail on habitable planet numbers and the nature of stellar dynamics in the Milky Way. Of course, we are far from a definitive answer, but astronomers are already working on the number of habitable exoplanets, and all of us are coming to terms with the challenges of civilizational longevity and planetary sustainability. In other words, we have to embrace a greener future before we can even talk about going boldly beyond.

The galactic archipelago model puts a new slant on the search for extraterrestrial intelligence. Rather than looking for individual exoplanets, we should search instead for those stellar archipelagos, those regions of the Galaxy where the topography of stars might promote interstellar planet hopping. And yet the main lesson to be learned from the model is this: it's totally normal for a planet like ours to have remained unvisited, as yet, by extraterrestrial species. In the same way that Pitcairn lay unexplored for centuries in the South Pacific seas, so Earth sits isolated in space, an island in waiting.

WHY ARE THEY CALLED STAR "SHIPS" AND SPACE "SHIPS"?

"For two thousand years men have believed that the Sun and the stars of heaven revolve around them. . . . The cities are narrow and so are men's minds. Superstition and plague. But now we say: because it is so, it will not remain so. I like to think that it all began with ships. Ever since men could remember they crept only along the coasts; then suddenly they left the coasts and sped straight out across the seas. On our old continent a rumor started: there are new continents! And since our ships have been sailing to them the word has gone round all the laughing continents that the vast, dreaded ocean is just a little pond. And a great desire has arisen to fathom the causes of all things. . . . For where belief has prevailed for a thousand years, doubt now prevails. All the world says: yes, that's written in books but now let us see for ourselves."
—Bertolt Brecht, *Life of Galileo* (1960)

VOYAGES OF DISCOVERY

In the *Star Trek* franchise, the starship is vital. These people-carrying spacecrafts are, after all, the very instruments of space exploration. They're capable of viable journeys to the stars. They transport crew and kit across galactic distances. And using propulsion tech, such as the warp drive, they are the key weapon of discovery of strange new worlds and civilizations. But why are they called space "ships"? Well, ships have been as vital to the development of modern science as they have to *Star Trek*. And the beginning of science reads something like a *Star Trek* backstory.

Modern terrestrial science began with ships. Two Chinese scientific inventions, the compass and the sternpost rudder, had a huge and global effect at sea. History was changed. Long voyages became possible. The seas were thrown open to exploration, a colossal expansion in trade, piracy, and war. The selfsame situation we see in *Star Trek*, only in a cosmic setting.

The need for better navigation had profound consequences for science. An open ocean demanded more accuracy: better observations, better charts, and better instruments. So open-sea navigation raised the need for a more predictive astronomy, a brand-new quantitative geography, and the desire for devices that could be used on board ship, as well as on land.

Scientific accuracy became so important that even pirates took notice. The pirates that preyed upon the high seas, a reflection of rival trade and colonization attempts by European powers, often sought a surprising booty. If a sea raid proved successful, the boarding pirates would head straight for the hold, for rather than gold, silver, or pieces of eight, the most precious cargo a ship possessed was its maps and chronometer. Indeed, some cartographers would knowingly include errors on their maps, to mislead the uninitiated should the map get into the hands of the wrong kind of pirate. (Kirk once remarked that Spock was a pirate at heart, due to Spock's counterpart from the mirror Universe fitting into such an environment so easily. One can easily imagine a Spock of the seven seas, logically heading for the hold and the maps and clocks!)

When Europeans started their piracy and plunder of the world, they needed to know where they were, and what they "owned." The scientific obsession with longitude began. Mariners for most of history had struggled with precise longitude. Latitude was easy enough; observing or predicting the positions of the Sun, Moon, planets, and stars would easily nail it. Mariner's tools such as the quadrant or astrolabe were used, reading off the inclination of the Sun or charted stars. With the great European sea voyages of the Portuguese explorers around 1415, the planet was opened up to capitalist enterprise. And these voyages were the fruit of the first conscious use of astronomy and geography in the pay of glory and profit.

SHIPS AND TELESCOPES AND STARS, OH MY!

The ship was of prime importance. For centuries, dry routes like the Silk Road had dominated. This was an old network of trade routes that linked the East and West. And it was crucial to economic, cultural, political, and religious life from the second century BC to the eighteenth century. After the oceangoing ship, all the major trade routes were wet. And now, economic power very much depended on the speed and consistency with which long-range voyages took place along those wet routes.

The European "discovery" of the Americas and the New World was the impetus for a more aggressive approach to nature. The quest for longitude and dominion in that New World promoted instrument making in science, just as the same quest of space would promote competition in *Star Trek*. And one of the first instruments of discovery on Earth was the ship itself.

Here's where the Star Trek backstory begins. The newly invented telescope was also a kind of ship. Consider this: pioneers of the telescope in the early 1600s took themselves and their contemporary world as onlookers to a place almost no one had ever imagined. And yet, like the destinations of a ship, the destinations of the telescope were also public knowledge. No one has put a better case for the ship as an instrument of discovery than twentieth-century German playwright Bertolt Brecht. In Brecht's greatest play, *Life of Galileo*, Tuscan astronomer Galileo declares from his "humble study in Padua" the quote at the start of this chapter.

Galileo's discoveries with the telescope from 1610 were so revolutionary it was as if an incendiary device had been hurled at the Church. Galileo used the telescope to dismantle the Church's idea of the Earth-centered cosmos. In their two-tier Universe, only the Earth was meant to change. Beyond the Earth, the celestial sphere was immutable and perfect. And yet through the spyglass, Galileo saw a picture that was dramatically different. The Moon was pockmarked and bumpy, not crystalline and perfect. And the lunar surface had Earth-like mountains, valleys, and hollows.

There was another way in which these discoveries were revolutionary: you could see for yourself. In the age of faith, science's enemy was monopoly. For centuries, knowledge had been shut up in the Latin Bible, which only black-coated ministers were allowed to interpret. Now, if you didn't believe what Galileo said of the lunar surface, anyone with enough ready cash to do so could pop out to the local spectacle-maker, pick up a spyglass, and see the surface of the Moon for themselves.

SAILS FOR SPACE!

The old cosmos was transformed. To the nine stars in the belt and sword of Orion, Galileo added eighty more; to the seven stars in the constellation of Pleiades, the seven sisters, he made thirty-six; and as for the belt of the Milky Way, the telescope resolved the Galaxy into a mass of innumerable stars seated together in clusters.

Now we begin to see how revolutionary was the telescope. If the magnificent splendor of the so-called God-given cosmos was made especially for our delight, why is it only through a newly invented machine, such as the spyglass, that we can savor its intricate parts, and begin to know its true nature? What's more, the telescope revealed a new and vast cosmos that stretched beyond the reach of human vision. It robbed the old Universe of its unity.

The old cosmos was transformed from the Moon outward. With its focal position in the night sky, the Moon was unique among the heavenly bodies. Its image through the telescope unveiled features that made it easy to describe by terrestrial analogy, such as "seas," "mountains," and so on. The Moon was evidence that there was material similarity between the Earth and the rest of the Universe. And the Moon acted as the basis for the first off-world science fiction stories exploring the possibility of alien life, such as *Somnium* (The Dream), a 1634 space voyage written by none other than the pope's mathematician, German genius Johannes Kepler.

After all, if we can find new worlds on Earth, like the Americas, why not also new worlds in space? Kepler's story, *Somnium*, is an extended voyage of discovery, from the oceans of Earth to the dark "seas" of the Moon. Galileo had inspired in Kepler early ideas about alien life. In his letter to Galileo, Kepler spoke about space travel and invented the spaceship:

> There will certainly be no lack of human pioneers when we have mastered the art of flight. Who would have thought that navigation across the vast ocean is less dangerous and quieter than in the narrow, threatening gulfs of the Adriatic, or the Baltic,

or the British straits? Let us create vessels and sails adjusted to the heavenly ether, and there will be plenty of people unafraid of the empty wastes. In the meantime, we shall prepare, for the brave sky-travelers, maps of the celestial bodies—I shall do it for the Moon, you Galileo, for Jupiter.

So, there we have it. Back in the days of Shakespeare, Galileo, and Kepler, the spaceship so ubiquitous in *Star Trek* was first invented, inspired by the telescope.

HOW DOES *STAR TREK* INTERPRET THE DRAKE EQUATION?

"[Carl] Sagan and [Iosif] Shklovsky had also speculated about guided evolution in Intelligent Life in the Universe, written in 1966. They based their argument on the Drake equation, a calculation of how many communicating alien civilizations there might be in our Milky Way Galaxy. Their reasoning went something like this: if there are sufficient technically accomplished alien civilizations out in the Galaxy, then some will look for contact with other less-developed planets, and every 100,000 years or so visit these planets to guide and monitor their evolution."
—Mark Brake, *The Science of Science Fiction* (2018)

A GALACTIC CHOWDER OF ALIENS

The Star Trek Galaxy is bursting with extraterrestrial life, from the apparent masters of the Universe in the Q Continuum to the free-loving Edo. The telepathic Betazoids, the mutable Changelings, the bulb-headed Talosians, the age-defying El-Aurians, and the ridiculous Tribbles—these *Star Trek* extraterrestrials present a wide range of alien intellect, charisma, and power. But is *Star Trek* realistic about the diversity of life in that fictional Galaxy? And does science offer us any way of estimating how much life there may be in our own Galaxy?

The answer lies in the so-called Drake Equation. This equation is the baby of US astronomer Frank Drake. In 1960, Drake had carried out the first modern search for alien life. He had pointed an eighty-five-foot radio telescope at two Sun-like stars in the solar neighborhood. The equation came later. It was written in 1961 and was conceived for a meeting at the National Radio Astronomy Observatory in Green Bank, West Virginia. At the meeting, Drake met up with a number of leading scientists in their fields. These guys knew their stuff. The team included legendary US astronomer and science writer Carl Sagan, Nobel Prize–winning chemist Melvin Calvin, renowned radio astronomer Otto Struve, talented neuroscientist John C. Lilly, and Manhattan Project physicist Philip Morrison. The science team, ten in all, called themselves The Order of the Dolphin, named after Lilly's work on dolphin communication.

THE DRAKE EQUATION

As *Star Trek* enthusiasts are at the hard-tech end of science fiction (here not be flying dragons, thank you very much) I shall assume readers are happy to take a mathematical plunge, so here's the Drake equation. It's composed of seven parameters, multiplied together, which comprise a set of seven limits on the number of communicating alien civilizations in our Milky Way. They are:

$$N = R^* \cdot f_p \cdot n_e \cdot f_l \cdot f_i \cdot f_c \cdot L$$

where:

N = the number of communicative civilizations in our Galaxy

R^* = the average rate of star formation in our Galaxy

f_p = the fraction of those stars that have planets

n_e = the average number of planets that can potentially support life

f_l = the fraction of planets that actually develop life

f_i = the fraction of planets with life on which intelligent civilizations arise

f_c = the fraction of civilizations that develop detectable signs of communication

L = the length of time such civilizations send communicative signals into the Galaxy

How does *Star Trek* interpret this equation? To consider this, and to understand how the number of communicating alien civilizations might be as large as in the Star Trek Galaxy, it's useful to calculate *Star Trek* values for each parameter and simply multiply them together.

Let's think about the nature of the parameters that make up the equation. The first five parameters (R^*, f_p, n_e, f_l, and f_i) are limits we can estimate using the logic of science and are mostly questions of quantity, with each of the f parameters usually expressed as a decimal between 0 and 1. Let's give some examples of estimates to guide us through. Astronomers can estimate the average rate of star formation in our Milky Way, and we are getting a better estimate of the fraction of stars in the Galaxy that have planets. The last two parameters (f_c and L) are concerned with human evolution and anthropology. Perfect parameters for *Star Trek*, as the rise and fall of planetary civilizations is the domain of speculative fiction rather than straight-up science.

PARSING THOSE PARAMETERS

Today, we have a far better idea of estimating Drake's first five parameters than we did when *Star Trek: The Original Series* first aired in 1966. Back then, before we even set foot on the Moon, science knew of no planets in orbit around other stars (though Bones was bold enough in "Balance of Terror" to maintain "In this Galaxy, there's a mathematical probability of three million Earth-type planets"). Indeed, even the entire run of *Star Trek: The Next Generation* came before the first exoplanets were discovered in 1995. But as of January 1, 2020, thanks to a number of ongoing science projects, and assuming there are 200 billion stars in the Milky Way, we can estimate that there are 11 billion potentially habitable Earth-sized planets in the Galaxy, rising to 40 billion if planets orbiting the numerous red dwarf stars are included. It seems Bones was too cautious in his estimation.

However, even though those first five parameters are questions of quantity, estimates can still vary wildly. Two astronomers, using the same data and same logical steps, can reach very different results. Astronomer one would conclude the data showed that *Star Trek* had got it right: the Milky Way really is full of alien civilizations. We simply haven't yet made contact. Whereas astronomer two would say that humans are alone in the

Universe. A bold conclusion, though, considering there are two trillion other Galaxies in the cosmos!

Remember that **N** is the number of communicating civilizations in just our Galaxy. **R*** is estimated to be a value of 1 by some astronomers, based on the current rate of star formation in the Milky Way. And yet the rate of star formation was once higher. So, dividing the number of stars in the Galaxy by its rough age of 10 billion years gives us a *Star Trek* value for **R*** of 10 per year, the number we shall plug into our Drake equation. Astronomers now have good estimates of f_p, the fraction of stars that have planets, so it's reasonably safe to plug in a value of 0.5 for f_p.

The next parameter is n_e: Is the planet ecologically fit for life? The answer is tied up with the idea of whether the planet orbits in the so-called Goldilocks Zone, where it would have a surface temperature just right to have liquid water. But it also might include factors such as having the type of atmosphere that stabilizes the planet's surface temperature, having a breathable atmosphere for its life-forms, and so on. Bearing all this in mind, let's enter a *Star Trek* value for n_e of 0.5.

The parameter f_l represents the fraction of planets that actually develop life. Earth seems primed for life, as fossil evidence of life appears in our rocks as long ago as 3.5 billion years. It's then easy to conclude, as *Star Trek* has, that life will form on planets wherever it can. Since some less optimistic scientists argue that life is very unlikely and f_l represents a bottleneck in the Drake Equation, we shall compromise and use a conservative value of 0.1 for f_l.

The parameters f_i and f_c represent whether planetary life begets intelligence and whether they ultimately communicate into space, respectively. If we join the writers of *Star Trek* and remain optimistic about Darwinian evolution and natural selection, then we can happily conclude that intelligence is powerfully adaptive and the value for f_i is 0.5.

As for f_c, the chance that the intelligent species can communicate through space, we shall not use a value of 1. You will recall that the original science team who worked on the Drake equation named themselves The Order of the Dolphin. Well, dolphins are truly intelligent but, being ocean-bound, they are unlikely to build powerful radio dishes that point at the sky—as far as we know! So, in the hope that around half of the planets which have evolved intelligent species can boast at least one dominant life-form that sends messages out into space, we can use an f_c value of 0.5.

Finally, we come to true *Star Trek* territory: **L**, the lifetime of a technological communicating civilization. Technical human civilization of this kind has been in existence for only a century. And for half of that, *Star Trek* has been busy telling stories about the exploration of space. What does the future hold for intelligent life on other planets? Would they become at least part cyborg? Or maybe an AI with regenerative capability? Or do so-called intelligent civilizations become dangerously fragile once weapons of mass destruction are created? Not to mention runaway growth ruining the home planet's ecology. The doubts about **L** overshadow all other parameters in the Drake Equation. Assuming most planets manage to avoid Armageddon, we shall adopt an optimistic Gene Roddenberry stance and assume a value for **L** of 4 million years.

Plugging all the above values into the seven parameters then generates the following answer:

$$N = 10 \bullet 0.5 \bullet 0.5 \bullet 0.1 \bullet 0.5 \bullet 0.5 \bullet 4 \times 10^6 = 250,000$$

In other words, the number of communicating civilizations in our Galaxy is a quarter of a million. Three things stand out from this calculation of ours. One, it supports the idea that our closest neighbors might be among the nearest stars. Two, when all the best guess data has been factored into the Drake parameters, the number of communicating civilizations in a Galaxy turns out almost equal to their average lifetime in years. Which means that the science fictional explorations of *Star Trek* have much to contribute to the debate. And three, the presence of alien civilizations in the Star Trek Galaxy really is based on a scientific footing.

TO BOLDLY GO: HOW HAS *STAR TREK* INFLUENCED SPACE CULTURE?

"Science fiction first emerged with science. Way back at the time of the scientific revolution, Earth became an alien planet. When scientists made the Earth-shattering discovery that we did not live at the center of the Universe, the revolution cut two ways. It made Earths of the planets, and it also brought the alien to Earth. Stories of space voyages helped us make sense of the nonhuman Universe in which we found ourselves: our marginal position in space, our fate in time, and the unravelling of the monster within us. Science fiction was a way of exploring the human meaning of scientific discoveries. It's a way of describing the cultural shock of discovering our marginal position in alien space; an attempt to put the stamp of humanity back onto the Universe; to make human what is alien."

—*Different Engines: How Science Drives Fiction and Fiction Drives Science* (2007)

"The newly unveiled logo for US Space Force appears to have boldly gone where *Star Trek* went before. Twitter users noted that the emblem, revealed by President Donald Trump, bears an uncanny likeness to the insignia from the cult sci-fi TV series. The striking resemblance left many critics as stunned as though they had been zapped by Captain Kirk's phaser."

—BBC News online (January 24, 2020)

THE SPACE AGE

The Space Age began with the launch of *Sputnik 1* on October 4, 1957, less than a decade before *Star Trek: The Original Series* first aired on September 6, 1966. Human civilization had passed through the Stone Age, the Bronze Age, and the Iron Age. The industrial revolution of the nineteenth century had been known as the Machine Age. Now, a new age dawned. And modern culture became increasingly catalyzed by the discoveries of space exploration. A perfect time, then, for *Star Trek* creator Gene Roddenberry to pitch the idea for a science fiction television series set in the twenty-third century aboard a starship whose crew's mission was to explore a fictional Galaxy.

A long-time fan of science fiction, Gene's first pitch on March 11, 1964, characterized the new show as "Wagon Train to the stars," a reference to a popular American Western television series that aired between 1957 and 1965. As the 1960s unfolded, real-life space exploration hit milestone after milestone. Following in Sputnik's wake, the Russian dog Laika became the first living creature in space. She was quickly followed by the first human in space, Yuri Gagarin in 1961, and the first woman, Valentina Tereshkova in 1963. In 1965, Alexei Leonov became the first human to carry out a spacewalk. Little wonder the first *Star Trek* crew included the Russian Chekov.

Once *The Original Series* was underway, humans quickly understood how to have a constant presence in orbit around the Earth and learned from the experience of being there. The Mir space station served as a microgravity research lab in which crews conducted experiments in biology, physics, astronomy, and meteorology. Research in spacecraft systems was also aimed at developing the technologies needed for the permanent occupation of space. Much later, the International Space Station became the product of lessons learned internationally, after decades of launching and operating such space stations. But let's wind the clock back a little. As the NBC network introduced *The Original Series* to its Fall 1966 program, NASA was getting ready to launch the Gemini 10 mission. NASA was almost three years from landing humans on the Moon. Meanwhile, the Starship Enterprise was tearing at warp speed through the fictional Galaxy.

STAR TREK SUCCESS

What were the key ingredients in *Star Trek*'s influential success? A fundamental factor was the portrayal of a more positive science fictional future. The show appeared particularly attractive to late 1960s television audiences due to the makeup of the Enterprise's crew which included a female African American communications officer, against the background of the civil rights movement; an Asian American senior helmsman, despite the recent wars waged on that continent; a Russian-born ensign, despite the then ongoing Cold War between the East and the West; and a half-human half-vulcan science officer, through whom the message became clearer still: our human/humanoid future in space will be best done through international, interracial, and even interstellar cooperation.

In 1976, NASA announced it would christen its first space shuttle Constitution, in homage of its rollout on September 17, the anniversary of the enactment of the US Constitution. *Star Trek* fans were successful in campaigning and persuading the shuttle name-change to Enterprise after President Gerald R. Ford directed NASA to rechristen America's first space shuttle. When the shuttle was unveiled in Palmdale, California, many of the original cast members of the show were present, as well as Gene Roddenberry. And so began a long relationship between the space agency and the Star Trek franchise.

Take the example of NASA's development of the Space Shuttle. During the 1970s, NASA needed to recruit a fresh tranche of astronauts to carry out systems operations associated with the Shuttle. Fly the vehicle, deploy the satellites, perform science experiments, that kind of thing. NASA also wanted to encourage women and minorities to apply to be astronauts, so the Agency hired Nichelle Nichols, the actor responsible for playing Lieutenant Nyota Uhura in *The Original Series*, to record a recruiting video in 1977 (for those interested, the video can be seen on YouTube, naturally!). The pitch paid off. In January 1978, when NASA announced the selection of thirty-five new astronauts, women and minorities were among the new recruits for the first time. Incidentally, on one of Ms. Nichols's return trips to NASA in September 2015, she flew aboard NASA's Stratospheric Observatory for Infrared Astronomy (SOFIA) airborne telescope aircraft. It's maybe the closest any of the cast members have come to actually flying in space.

James Doohan, the actor who played Lieutenant Commander Montgomery "Scotty" Scott, the Starship Enterprise's chief engineer, has also had a number of associations with NASA. Most notably the unique tribute at Doohan's retirement in 2004 when none other than astronaut Neil Armstrong, the first human to step out onto the Moon, made a rare public appearance to speak as "one old engineer to another."

TREK SUCCESS

As well as *Star Trek* fans being instrumental in naming the first Space Shuttle, NASA felt the franchise's influence in other ways. When the space agency was in the design stage for an Earth-observation facility for the International Space Station (ISS), the optical quality window, with the formal designation of Window Observational Research Facility, quickly became WORF for short. The link between fact and fiction, between research facility and the Klingon officer of the Enterprise in *Star Trek: The Next Generation* was a chance no one wanted to miss. NASA added a further detail to ensure everyone got the point: On the research facility's official patch, below the acronym WORF, is the same name written in Klingon.

A number of NASA crews have adopted *Star Trek* tropes for their unofficial photographs. For example, the crew of STS-54, Space Shuttle *Endeavour*, was decked out in the costumes of Starship Enterprise officers from *Star Trek II: The Wrath of Khan*. And when Space Shuttle and ISS crews create Space Flight Awareness posters for their respective missions, on more than one occasion they have embraced *Star Trek* motifs. ISS Expedition 21 chose to deck out in costumes from *The Original Series*, while the Space Shuttle Endeavour STS-134 crew opted for uniforms from the 2009 reboot motion picture *Star Trek*.

It's clear that science and *Star Trek* go hand in hand. *Star Trek* has had a significant cultural influence on not only scientists and engineers, but even astronauts. Sadly, contemporary spacecraft are yet to speed through our Galaxy like the Enterprise, but the franchise's inspirational influence makes progress to that end a more likely reality.

TO BOLDLY GO

There's one final way in which contemporary space culture feels the strong influence of *Star Trek*. Larry Niven, US science fiction author, once said that the dinosaurs became extinct because they didn't have a space program. Larry was being quoted by Arthur C. Clarke, the author of *2001: A Space Odyssey*. Clarke had been in conversation with Buzz Aldrin, the second man to walk on the Moon. The two futurists were contemplating our real space odysseys in the century that lies ahead. Larry's point was this: If the dinosaurs had had a base on the Moon or Mars, they'd have had a better chance of survival. Those raptors were smart.

One astronaut who's taken the *Star Trek* philosophy on board is Jeff Hoffman. Hoffman wishes to see us off-Earth in our lifetime. He believes the technology to reach nearby planets is already there. Hoffman is one of only five hundred humans who have viewed the Earth from orbit. You know the perspective this experience affords: while in orbit, Hoffman could put his hand up to his eyes and block out our planet. All we know, all of humanity, and all of human culture essentially disappears from view. This perspective shows how vulnerable humans are. So, for Hoffman as well as Niven, the long-term survival of our species depends on us becoming a multi-planet being. Better that than ending up like the raptors.

Space really isn't very far at all. Space begins a mere 62 miles away. In the grand scheme of things, that isn't very far. If you are in London, you can literally say that space is closer than Paris. Or, if you live in Seattle, Canberra, Hyderabad, Cairo, Beijing, or central Japan, space would be closer to you than the sea. Naturally, we need to expend energy to get there. But it doesn't have to be so expensive, and we are beginning to see the start of space transportation that could be the leading edge of a revolution.

A lot of the social forces here on Earth might fundamentally change when we are living on other worlds. Humans could set up new kinds of social structures and new kinds of philosophy and politics. Exactly those kinds of scenarios that *Star Trek* has explored over the last fifty years or so. It would be an enormously revolutionary step, not just in human evolution but also an evolution in living.

Given our future on a threatened planet, it might be wise to take our inspiration from *Star Trek*. To boldly go, to colonize and spread ourselves throughout the Galaxy, or at least make a start, so we are not as prone to a single catastrophe such as a stray comet or an attack from The Borg. NASA may not yet have warp drive. But we humans need to continue doing what we've done for over half a century—sending missions out into the solar system and beyond, to seek and explore new worlds.

WHAT CAN *STAR TREK* TELL US ABOUT PLANET HUNTING?

"The American space observatory Kepler, launched in 2009 to find Earth-like planets orbiting other stars, took off four hundred years after Galileo's first use of the telescope, and is of course named after the first great Copernican theorist, Johannes Kepler. Based on Kepler's early findings, Seth Shostak, senior astronomer at the SETI institute, estimated that "within a thousand light-years of Earth," there are, "at least 30,000 habitable planets." And based on the same findings, the Kepler team projected that there are "at least 50 billion planets in the Milky Way" of which "at least 500 million" are in the habitable zone. NASA's Jet Propulsion Laboratory was of a similar opinion. JPL reported an expectation of two billion 'Earth analogues' in our Galaxy, and noted there are potentially around one sextillion Earth analogue planets."

—Mark Brake, *Alien Life Imagined* (2012)

CLASS M PLANETS

In *The Next Generation* episode "Justice," the Enterprise arrives for shore leave at the idyllic and newly discovered planet of Rubicun III. A small party from the Enterprise is sent down to meet with the Edo, the curiously frisky native people of the planet. As Picard describes it in his log, "Stardate 41255.6. After delivering a party of Earth colonists to the Strnad solar system, we have discovered another Class M planet in the adjoining Rubicun star system."

What exactly is a Class M planet, and does the idea have any parallels in astronomy proper? In the system of planetary classification that the Federation used, a class M planet or Moon was a world thought to be suitable for humanoid life. By the twenty-fourth century, thousands of Class M planets had been charted by the Federation. You might also recall that, in the *Star Trek: First Contact* movie, when Lily asks Picard how many planets are in the Federation, he replies that there are "over one hundred and fifty, spread across eight thousand light-years" of space.

Class M worlds were the first choice for colonization. Beginning in the twenty-second century, humans, and later the Federation, terraformed lifeless planets like Mars into Class M worlds. Witness the examples from "The Cage" in *The Original Series*, "Caretaker" in *Voyager*, as well as *The Next Generation*'s "Justice," "Home Soil," "Final Mission," and "Haven." Furthermore, conditions on board Federation starships mimicked the Class M environment for maximum comfort.

EXOPLANET HUNTING

Like the Federation, astronomers are busy charting planets beyond our solar system. In fact, we live in a great age of planetary discovery. The first confirmation of a so-called exoplanet orbiting an ordinary star was made in 1995, almost thirty years after *The Original Series* and barely twelve months after *Star Trek: Generations* hit the cinemas. The first exoplanet discovery was a giant planet moving in a four-day orbit around the nearby star 51 Pegasi. On October 6, 1995, Swiss astronomers Michel Mayor and Didier Queloz announced their discovery and on October 8, 2019, Mayor and Queloz shared the Nobel Prize in Physics for this same discovery.

Rather than being called Class M planets, astronomers refer to such worlds as Earth-like planets, or sometimes Earth "analogues." Since 1995, astronomers hunting for potentially life-bearing Earth-like planets around Sun-like stars reckon there may be tens of billions in our Galaxy alone. A European team of scientists reported that perhaps 40 percent of the estimated 160 billion red dwarfs in the Milky Way have a "super Earth" orbiting at a distance that would allow water to flow freely on its surface.

But first, how can we be sure that such exoplanets exist? As exoplanets are very faint points of light, compared to their Suns (usually around a million times fainter, in fact), looking for them would be like looking for a needle in a haystack. So, only a very few exoplanets have actually been seen. For the rest of these extra-solar worlds, astronomers have to use other methods to find them. And one such way is to look for what we might call wobbly stars.

Consider a candidate star, wheeling its way through the Galaxy. Though the star is huge, even compared to our Sun, it's not as stable as it seems. A star with planets in orbit about it will move in a slightly wobbly orbit because of the effect of the gravity of the attendant planets, so when planet hunters see a wobbly star, they know there may well be a planet in tow. And by measuring the size of the wobble, the hunters can estimate the size of the planet.

Imagine Vulcan planet hunters out in space. Not on the planet Vulcan itself but sailing along in one of the Vulcan survey ships used by the Vulcan High Command from the mid-twentieth to twenty-second centuries. If they were to study our solar system, they would see a wobbly Sun. Like all stars with planets, our Sun wobbles. The combined gravity of the planets is what causes the Sun's wobble. But, as 99 percent of the mass of our system which lies outside the Sun resides in mighty Jupiter, the Sun's wobble is mostly down to our system's giant planet.

So far, looking for wobbly stars is one of the best ways of finding exoplanets. Thousands have been discovered, and many more await discovery. Terran planet hunters look for twitchy stars out to about 160 light-years from Earth. The "Jupiter effect" means they tend to find very large planets in orbit about wobbling stars. That's because bigger planets mean bigger wobbles, and that means this way of planet hunting is not so good for finding other Earths, the Class M worlds that feature in *Star Trek*.

EXOPLANET HOLY GRAIL

For Terran planet hunters, their "Holy Grail" is to find other Earths, our Class M equivalents. These exoplanets are naturally much smaller than gas giant planets like Jupiter, and so more difficult to detect. In our solar system, a planet must sit at a certain distance from the Sun if it is to support life. If a planet is beyond the outer limits of this zone, it won't get enough of the Sun's energy, and water will freeze. If a planet is within the inner limits of this zone, it will get too much solar energy, and surface water will "boil" away. This not-too-hot, not-too-cold idea is why astronomers sometimes call this zone the Goldilocks Zone.

Humans have wondered for centuries if there are stars out in deep space with planets in orbit about them. Relatively recent research has convinced astronomers there may be many billion such worlds, many billion such Class M equivalents. For example, in November 2013, based on space mission data, astronomers reported there could be as many as 40 billion Earth-sized planets orbiting in the Goldilocks Zones of Sun-like stars and red dwarfs in the Milky Way. And, of these, 11 billion may be orbiting Sun-like stars. 11 billion! No wonder Arthur C. Clarke wrote in *2001*, "there is enough land in the sky to give every member of the human species, back to the first ape-man, [their] own private, world-sized heaven—or hell."

It seems *Star Trek* was right about Class M planets. They're out there in almost unimaginable numbers. Science fiction writers have, if anything, been somewhat conservative about how replete space is with potentially habitable planets. A recent review suggested that exoplanets Kepler-62f, Kepler-186f, and Kepler-442b are the best candidates for being potentially habitable. These worlds are at distances of 1,200, 490, and 1,120 light-years away from Earth, respectively. Of these, Kepler-186f is similar in size to Earth with a 1.2-Earth-radius measure and it is located toward the outer edge of the habitable zone around its red dwarf. The potentially habitable planet TOI-700 d is only 100 light-years away. All we need now is a workable warp drive.

ARE *STAR TREK* AND PICARD RIGHT ABOUT THE TERRAFORMING OF MARS?

"Mars's soil must be transformed, or 'terraformed.' Special units would pump gases, such as methane and ammonia, into the atmosphere. The gases would absorb solar energy and warm the planet, triggering the release of carbon dioxide from the soil and ice caps. The carbon dioxide in the atmosphere would help global warming, so oceans would form. After several decades of terraforming, Mars might look as blue and watery as Earth. Within a century, it could be terraformed into an oxygen-rich environment, supporting a human colony, some of whom may dream of traveling to the remote corners of the solar system, and beyond."

—Mark Brake, *How to Be a Space Explorer* (2014)

"I THINK IT'S THE BEST JOB IN THE UNIVERSE."

In the *Star Trek: Picard* episode "Remembrance," we are reminded that on April 5, 2385, Mars was attacked by rogue synths. They destroyed the Utopia Planitia shipyards and bombarded the Red Planet so heavily that flammable vapors in its stratosphere were ignited. Mars continued to burn well into 2399. In the *Star Trek* timeline, Mars was the first planet terraformed by humans. Colonists originally dwelt in domed cities, but by

2155 the Martian lowlands were sufficiently engineered to allow humans to freely roam without suits.

The question of terraforming plays a central role in the *Next Generation* episode "Home Soil." Diverted from exploring the Pleiades, a constellation first observed through the telescope by Galileo, the Enterprise arrives at the terraforming colony on Velara III, where they meet terraforming engineers, including Luisa. She explains how they take a lifeless planet and transform it little by little into a Class M environment capable of supporting life. Not only does Luisa declare hers to be "the best job in the Universe," she also admits that terraforming "makes you feel a little godlike." For the benefit of the viewers, she explains that the first phase involves selecting the planet. It must have the right mass and gravity, the correct rate of rotation, and a balanced day and night. (Sounds like our own Mars would be a suitable candidate for this.) She says that the planet must also be without life, or the prospect of life developing naturally. (Mars qualifies here too. Scientists have not yet detected life on Mars, though there is speculation that ancient Mars once had running water, so it may have had primitive life-forms in the past.) Ultimately, Luisa says, the terraformers engineer the planet's water content (filtering the water, removing the salt, oxygenating, and replacing) until they can introduce micro-organisms and set the scene for "a lush, arable biosphere." Will terraforming be this easy in reality? And would you wave goodbye to Earth and live out the rest of your days on Mars if terraforming was that easy?

TERRAFORMING THE RED PLANET

Mars, our prime candidate for terraforming, certainly qualifies in some regards. It's close to home, it has seasons of a sort, it rotates in 24 hours, it's tipped on its axis, it has a balanced day and night, and it has polar ice caps. And yet in other ways, Mars is not so homely. It has a mean temperature of -81°F and an atmosphere so thin that if you stood on it unprotected, your saliva would simply boil away.

But many people, including tech entrepreneur Elon Musk, hold out hope that one day a colony of humans will live on Mars. The Red Planet is the obvious escape plan, a place to go if Earth gets hit by some celestial body, suffers nuclear destruction or some pandemic, and assuming we don't meet up with another rogue synth Mars attack. In short, if humanity wants to become a proper two-planet species, terraforming Mars would help a lot, even if it's only a convenient observation platform from which to watch the end times on Earth.

As Luisa suggests, the first step toward making Mars our second home would be raising the planet's temperature and atmospheric pressure enough for living organisms to survive. But the Martian atmosphere is very thin. And pressure on the planet is so low that water cannot exist at the surface in liquid form. If you, dear reader, were to be dropped onto the Martian surface right now, you wouldn't have enough oxygen to breathe and, like Mark Watney in the 2015 movie *The Martian*, you'd have to wear a pressurized suit. To create an atmosphere similar in composition and thickness to that of Earth, carbon dioxide (CO_2) and other greenhouse gases would be needed to warm up the Red Planet.

These greenhouse gases help trap heat in a planet's atmosphere. While it's true that there's too much warming of the Earth's atmosphere right now, a little bit of heating would totally transform Mars. Planet Earth's blanket of gases protects terrestrial life from the Sun's radiation and keeps our climate livable. Some past research suggested that there may be enough CO_2 locked away in the Martian surface to heat the Red Planet and thicken the atmosphere. If the CO_2 can be released.

THE TROUBLE WITH TERRAFORMING

The problem is that data from the spacecraft and rovers monitoring Mars suggests there may not be enough CO_2 in the Red Planet's possible reservoirs. In short, and crucially taking into account the continuous leaking of Martian CO_2 into space, there's not enough greenhouse gas to warm the planet. The best guess is that the CO_2 reservoirs would merely triple Mars's atmospheric pressure—that's only one fiftieth of the change needed to make Mars livable. That amount of CO_2 would increase the Martian surface temperature by less than 50°F.

Neil deGrasse Tyson's solution to this is what he calls B-Y-O-G-H-G: Bring Your Own GreenHouse Gas! Tyson has also explained Elon Musk's plan of putting in Mars orbit satellites that have big reflectors, which focus Sunlight that would otherwise miss the planet. The reflected Sunlight would add more energy to the planet, heating it up, and help evaporate more CO_2.

As well as the Red Planet not supporting a thick enough atmosphere for humans, there is another Martian challenge: Mars doesn't have a sufficiently strong magnetic field to properly protect life. Earth's molten core creates a magnetic field around our planet that helps protect the terrestrial atmosphere from harmful solar radiation. Instead, rays from the Sun are deflected by the magnetic field, so they don't strongly impact our atmosphere and damage it.

Planetary scientists think that Mars once had a molten core and a magnetic field but both were lost billions of years ago. The Martian surface is now unprotected from the solar wind—the torrent of charged particles that stream from the Sun's interior and out into space. The upshot is that the gases in the thin Martian atmosphere are forever leaking into space. Indeed, research from recent missions to Mars suggests that the majority of Mars's ancient and potentially habitable atmosphere has been lost to space, torn away by perpetual solar wind and radiation, and never to return. NASA summed up the findings of their Mars Atmosphere and Volatile Evolution (MAVEN) mission by saying, "The solar wind is a stream of particles, mainly protons and electrons, flowing from the Sun's atmosphere at a speed of about one million miles per hour. The magnetic field carried by the solar wind as it flows past Mars can generate an electric field, much as a turbine on Earth can be used to generate electricity. This electric field accelerates electrically charged gas atoms, called ions, in Mars's upper atmosphere and shoots them into space."

A MARS A DAY HELPS YOU WORK, REST, AND PLAY

A crucial component of terraforming Mars is to protect its atmosphere (present and future-built alike) from being lost to space. According to Japanese scientists Osamu Motojima and Nagato Yanagi from the National Institute for Fusion Science, a planet-wide artificial magnetosphere could be created, which would solve the problem. The build, already feasible using current tech, would be a system of planet-encircling superconducting rings, each carrying a sufficient amount of direct current electricity.

Another study suggests deploying a magnetic dipole shield at the Mars L1 Lagrange point. Lagrange points are positions in space where the gravitational forces of a two-body system, like the Sun and Mars, produce enhanced regions of attraction and repulsion. Such points can be used to reduce fuel consumption needed to remain in position. A Mars dipole shield, creating an artificial magnetosphere between Mars and the Sun, would protect the Red Planet from solar wind and radiation. Simulations suggest that, within a matter of mere years, Mars would achieve half of Earth's atmospheric pressure. Without solar winds stripping away at the Martian atmosphere, frozen CO_2 at the ice caps would change from solid into gas and warm the equator. Polar melt would in time form seas, enough to fill one-seventh of Mars's prehistoric oceans. If all this succeeds, and Mars is terraformed, someone should keep a close eye on those rogue synths in the future!

HOW DOES *STAR TREK* PORTRAY ALIENS?

"But who shall dwell in these worlds if they be inhabited? Are we or they Lords of the World? And how are all things made for man?"

—Johannes Kepler, quoted in *The Anatomy of Melancholy* (1621)

"Why would a vastly superior race bother to harm or destroy us? If an intelligent ant suddenly traced a message in the sand at my feet reading, "I am sentient; let's talk things over," I doubt very much that I would rush to grind him under my heel. Even if they weren't super intelligent, though, but merely more advanced than mankind, I would tend to lean more toward benevolence, or at least indifference, theory."

—Stanley Kubrick, *Playboy* interview (1968)

IMAGINING THE UNIMAGINABLE

How does *Star Trek* imagine the unimaginable when it comes to alien life? Given that it's hard to imagine what is unknown, how does the franchise go about representing the unimaginable? On what science fiction traditions do such representations rely? For most of civilized history, humans have been unconcerned with the possibility of life beyond this small world on which we live. This planet. Indeed, for most of civilized history, we weren't even aware of the fact that we lived on a planet at all.

Then, when the Scientific Revolution began in the seventeenth century and we were able to train our telescopes skyward, human imagination became a tool to visualize the unknown. A tool to imagine potential worlds, a tool to imagine other Earths, and a tool to imagine alien life. Writers, in particular, searched for ways in which to push the boundaries of the imagination and expand the outer limits it faced in imagining extraterrestrial life.

For most of the last century or so, ever since H. G. Wells's 1898 novel *The War of the Worlds*, science fiction writers and movie directors thought long and hard about their portrayal of creatures from other worlds. In that time, some other alien archetypes have evolved. The predatory and possessive alien mother in Ridley Scott's 1979 movie *Alien*. The swirling sentient sea in Stanislaw Lem's 1961 novel *Solaris*, later made into movies in 1972 and 2002. And the hyperactive glove puppet with poor sentence construction from George Lucas's Star Wars movie series.

IT ALL CHANGED AFTER DARWIN AND WALLACE

Alien as highly evolved killer, alien as ocean-planet, and alien as wise, benevolent, if slightly ridiculous, mentor. But the portrayal of aliens, in fact or fiction, wasn't always so varied. Before science fiction really began in the nineteenth century, aliens were not genuine extraterrestrial beings. They were merely men or animals living on other Earths. It was Charles Darwin and Alfred Russel Wallace who changed all that, for Darwin and Wallace invented the alien. Their theory of evolution gave science fiction at least some basis for imagining how life might develop beyond this planet. From now on, the notion of alien life was linked with the physical and mental characteristics of the true extraterrestrial. And the idea of the alien became deeply embedded in the public imagination.

The archetypal alien, with its strange physiology and intellect, owes a huge amount to *War of the Worlds*. Wells's martians are agents of the void. They are the brutal natural force of evolution, and history's first menace from space. Wells's genocidal invaders, would-be colonists of planet Earth, were so influential that the alien as monster became a cliché in twentieth-century science fiction.

WELLS'S MARTIAN IN THE FRANCHISE

Wells's alien eventually found its way into *Star Trek*. Consider the following examples. The bipedal reptilian race known as the Gorn made its first appearance in *The Original Series*. The Gorn is very deadly due to a heady mix of superstrength, physical durability, and brutal weaponry like the advanced tech of Wells's martians, whose superiority means that the bright future of Victorian society is abruptly blown apart by the natural force of evolution in the shape of the Martian attack of Earth.

The same might be said for the Tholians. Appearing in a number of different series within the franchise, the Tholians are not only hostile and territorial in nature, though admittedly Wells's martians invade Earth because Mars is dying, the Tholians also have unique forms of tech and bio-weaponry to add to their CV. (One such weapon was the so-called Tholian Web, an energy field that could trap starships, entrapping them in the knowledge that any attempt to leave would result in the ship being shredded by the filaments.) Unlike the martians, however, the Tholians are not an expansionist race.

We also find reflections of Ridley Scott's highly evolved killer alien in *Star Trek*. Most notably in Species 8472, the only entity capable of striking fear into the Borg. Known solely by their species serial number assigned by the Borg, Species 8472 have an extremely resilient genetic programming which makes them immune to assimilation. No wonder the Borg see Species 8472 as a genetically perfect race. Highly evolved killer, indeed, and worthy of extreme admiration and terror.

Wells's book famously begins, "across the gulf of space, minds that are to our minds as ours are to those of the beasts that perish, intellects vast and cool and unsympathetic, regarded this Earth with envious eyes, and slowly and surely drew their plans against us." If, as we said, the martians are a force of nature, of evolution, and agents of the void, can we say the same for Species 8472? Very much so. The Federation might have penetrated space through discovery, but now they've more than met their match. Like Wells's martians, Species 8472 are a menace from space so profound that they convey the quality of the void—immenseness, coldness, and indifference—in their rendering. It seems that all attempts at resistance are futile, furthering the idea of Species 8472 as an unrelenting force of deep space.

WISE AND BENEVOLENT MENTOR ALIENS IN THE FRANCHISE

But what of the wise and benevolent mentor type of alien in *Star Trek*? The Cytherians were a species who knew subspace distortion so well that they could implant their consciousness into spaceships as a means of driving it to their home planet. Rather wisely, this meant that the Cytherians could pick and choose who they made contact with, instead of just chance encounters with various other species (you never know when you might run into the Borg or Species 8472, after all).

The Founder Humanoids, who appear in *Star Trek: The Next Generation* episode "The Chase," are responsible for seeding multiple planets to spread life throughout the fictional Galaxy. As the oldest and first race to obtain interstellar travel, they found the Galaxy bereft of life, and so populated it in their image, which surely gives them some kind of godlike status. Such a status could also be assigned to those non-corporeal blue and translucent orbs of energy known as the "wormhole aliens," or the Prophets. Allegedly living within a stable wormhole, the Prophets moved in mysterious ways, meeting other species by either possessing their body or appearing to them in a dreamlike state of someone that individual knew.

ELUSIVE ALIENS

These portrayals of non-corporeal aliens are interesting ones. When we spoke earlier about the way in which science fiction like *Star Trek* tries to imagine the unimaginable when it comes to alien life, in the tradition of writers such as H. G. Wells, we should also note that, in all that time, science has had little to say about the actual details of extraterrestrial life. We've still not found ET. In the face of a lack of hard evidence, and in an attempt to reinvent itself, some writers and directors have turned to more elusive explorations of alien contact.

Perhaps the most celebrated of these portrayals is Arthur C. Clarke and Stanley Kubrick's 1968 movie *2001: A Space Odyssey*. Kubrick claimed the picture provided a "scientific definition of God" by tracing the evolution of humans, from ape to superman, through the agency of the episodic guiding hand of superior alien beings.

In 1968, Kubrick did a fascinating interview with *Playboy* in which he talked about the kinds of issues artists face when trying to imagine the unimaginable. He declared that he didn't believe in any of Earth's mono-theistic religions, but he did believe that one can construct an intriguing scientific definition of God. "When you think of the giant technological strides that man has made in a few millennia, less than a microsecond in the chronology of the Universe, can you imagine the evolutionary development that much older life-forms have taken? They may have progressed from biological species, which are fragile shells for the mind at best, into immortal machine entities—and then, over innumerable eons, they could emerge from the chrysalis of matter transformed into beings of pure energy and spirit. Their potentialities would be limitless and their intelligence ungraspable by humans."

THE Q

Kubrick appears to be speaking here not just about his own wise and benevolent aliens in *2001*, but also of the similar generic type that appear in *Star Trek*. Let's compare Kubrick's own portrayal of superior aliens with the immortal and extra-dimensional alien race in *Star Trek* known as the Q, probably the most advanced extraterrestrials ever to be seen on the franchise. Their "enviable" abilities include knowledge of how to tamper with time, as well as the mastery over any form of matter in the cosmos.

We first meet the Q in "Encounter at Farpoint," the pilot episode and series premiere of *Star Trek: The Next Generation*. They are initially portrayed as a cosmic entity, judging humanity to see if it is becoming a threat to the cosmos. And yet, as the seasons progress, their role becomes more that of a teacher to Picard and humanoids generally, as the crews of Starfleet vessels are enlightened as "inferior" races. The Q are shown to be capable of matter-energy transformation and instant teleportation to anywhere, anyplace, anytime, along with possessing intellects so vast that many Q claim to be all-knowing.

And yet, compared to Kubrick's aliens, the Q often enlighten "inferior" races in quite destructive and disruptive ways, seemingly subject to their own whim and amusement. Remember that Clarke and Kubrick's *2001* is an epic journey. German philosopher Friedrich Nietzsche identified three stages in the evolution of man: ape, modern man, and ultimately, superman. And Kubrick's motion picture traces man's journey through these three stages. As the movie's subtitle suggests, the narrative is a spatial odyssey from the subhuman ape to the post-human star child. How is this evolution achieved? Through the guiding hand of an elusive alien intelligence that we never see. American astronomer and SETI pioneer Carl Sagan congratulated Kubrick's decision to portray the *2001* aliens not as humanoid but as an extraterrestrial superintelligence, fitting for the Nietzschean theme of man's evolving into post-human superman. When pressed by *Playboy* in 1968, Kubrick hinted at the nature of the elusive aliens in *2001* by suggesting, given their long maturation, that they had evolved from biological beings into "immortal machine entities" and then into "beings of pure energy and spirit," beings with "limitless capabilities and ungraspable intelligence."

The Q also give humanity a lesson (somewhat like Kubrick's elusive and wise aliens), albeit playful, if not ridiculous. When the main Q character first appears on the bridge of the Enterprise, he calls out humanity for their war-like and destructive nature, putting humans on trial and forcing Picard to serve as counsel for the defense. Q then reappears several times throughout the seasons, often in silly episodes, such as "Qpid," where Enterprise is sent to Sherwood Forest for a Robin Hood adventure. By the time of the finale "All Good Things," Q reveals that humanity's trial never ended. In short, the whole series was Q's trial as the crew learned and grew from their many interactions with alien races, hostile or friendly.

PART II
TIME

THE INNER LIGHT: HOW MIGHT ALIEN CIVILIZATIONS LEAVE FOOTPRINTS IN TIME?

"The activity of a highly developed society of intelligent beings could have cosmic significance and could produce artifacts which would outlive the civilizations that constructed them."
—Carl Sagan and Iosif Shklovsky,
Intelligent Life in the Universe (1966)

THE INNER LIGHT

The Next Generation episode "The Inner Light" is regarded by many critics and fans alike as one of the very best in the franchise. The Enterprise finds a mysterious probe that is seen to swiftly scan the ship and focus an energy beam on Picard, who appears to wake and find himself on a planet called Kataan. Family and friends convince Picard that he is in fact Kamin, and that his memories and true identity are mere dreams. Picard adjusts as the years seem to pass and he begins to notice that the planet Kataan is suffering a drought due to increased radiation from the local Sun.

Knowing that Kataan is doomed, the planet's elders place memories of their society into a probe. The probe is to become an abiding artifact of their planetary culture and is launched into space. Someday, they hope, someone from another civilization will find the probe and the Kataani (if we may call them that) will live on in memory. Picard, as Kamin, suddenly comes to a realization, *Oh, it's me, isn't it? I'm the someone . . . I'm the one it finds.*

We see Picard wake up on the bridge of the Enterprise. He discovers that, though he had felt he'd lived a lifetime on Kataan, in fact only twenty-five minutes have passed. The probe ends its beamed broadcast and is taken aboard the Enterprise. Inside the probe, the crew discovers a small box, which Picard opens to find a flute that he had learned to play during his life as Kamin. The episode ends as Picard picks out a poignant piece on the flute, a musical eulogy to a dead, but no longer forgotten, planetary civilization.

INTELLIGENT LIFE IN THE UNIVERSE

Just over a quarter of a century before "The Inner Light" was broadcast, American space scientist Carl Sagan published his famous book *Intelligent Life in the Universe*, written with Soviet astrophysicist Iosif Shklovsky. This book was one of the first serious attempts not just to cover the natural evolution of the Universe, but also to try engaging in scientific speculation about the development of intelligent civilizations among potential galactic societies. It has a special place in the history of science fiction and served as Stanley Kubrick's scientific "bible" while researching and filming arguably the twentieth century's greatest science fiction movie, *2001: A Space Odyssey*.

In the book's final chapter, the authors consider intelligent alien civilizations. They discuss the possibility that the lifetime of a civilization is not indefinitely long. And they stress that the death of civilization on one small planet does not imply the end of intelligent life in the Universe. This also appears to be the ethos the Kataani adopt in "The Inner Light." We can gather that, at the time of its passing, the Kataani culture was on the brink of some form of post-industrial society. We see no obvious evidence of heavy industry. The citizens enjoy an agrarian lifestyle. And yet their culture was sufficiently evolved to launch interstellar probes. As Sagan and Shklovskii point out, just as one individual can introduce some concrete, if small, contribution to society, a given planetary culture may make a contribution to the evolution of intelligent life in the Universe. Furthermore, just as the contribution of an individual in society would be nothing without its communication, the cultural contribution of a planet to the evolution of intelligent cosmic life simply cannot happen without interstellar communication. Thus, the Kataani probe.

In the book's section on interstellar contact by automatic probe vehicles, the authors discuss the interesting possibilities of physical alien contact through relatively short-range interstellar probes. How physical objects, like the Kataani flute, could be transported to civilizations in the stellar neighborhood; how the likelihood of encountering them far from their source would be small; and how the artifact diffusion might connect civilizations of differing levels of technical culture. Finally, they also speculate how the lifetime of the probe could be much longer than the lifetime of the civilization which built it. How the probe might dwell in space for millions of years after the parent civilization had perished. And how such a probe might transmit the heritage treasures of the culture of a dead civilization into the cosmos for hundreds of millions of years. Such as it was with the Kataani.

SLAVE TO THE RHYTHM

The other key element to the Kataani cosmic plan is their tampering with Picard's time perception. After being hit by the radiating beam from the Kataani probe, Picard lives a lifetime on Kataan as Kamin, and yet barely half an hour elapses on the Enterprise. The science of time perception is an academic field that involves psychology, neuroscience, and cognitive linguistics. Its main subject of study is the way in which the subjective experience of time is measured by someone's consciousness. Although directly measuring another human's time perception is not possible at the moment, scholars can study and infer perception through a number of experiments that help expose the underlying neural mechanisms of time perception. There are a number of theories for modeling time perception in the brain, each with limited success in terms of evidence in brain physiology or anatomy. However, one interesting philosophical theory of time duration came from Rabbi Nachman of Breslov who suggested that only the present day and present moment are "real." Interestingly, with regard to the plot of "The Inner Light," Rabbi Nachman also said that a person could sleep for a quarter of an hour and yet dream that they had lived seventy years.

Time perception is not associated with a particular human sensory system. But neuroscientists suggest that humans do indeed have a system, or perhaps several complementary systems, governing time perception.

One such component is the suprachiasmatic nucleus, a tiny region of the brain in the hypothalamus. The nucleus governs the circadian (or daily) rhythm, while other cell clusters may govern shorter (ultradian) timekeeping. There is also some evidence that very short (millisecond) durations are processed by slave neurons in early sensory parts of the brain.

You may have experienced altered time perception. For example, neuroscience research suggests that time appears to slow down for a person during dangerous events, such as a car accident, or a robbery. At these moments, the people experiencing the time dilation seem capable of complex thoughts in what would normally be the blink of an eye. These accounts of the apparent slowing of temporal perception may have been advantageous in our evolutionary past, especially at those times when our ability to intelligibly make quick decisions was of crucial importance to our survival. (To be clear, observers very often report that time appears to have moved in slow motion during such events. But it's not clear whether the experience is a function of increased time resolution during the actual event, or rather an illusion created in retrospect, by the recall of an emotionally memorable event.)

Given what we already know of time perception, it seems reasonable to assume that an advanced civilization might be able to tweak perception to some degree, though admittedly we don't know just how advanced the Kataani were, and this is some ambitious tweak they perform on Picard!

HUMANITY'S OWN INNER LIGHT

We humans have sent out our own version of the Kataani probe: two ships, launched into the huge dark ocean of deep space. Their journey has been the longest odyssey in human history. Sailing through space at around 40,000 miles an hour, the ships rounded Jupiter, using its gargantuan gravity as a slingshot, and were flung out of the solar system to journey through the Galaxy for a billion years.

A small team was assembled to create the ships' message. One that would last something close to eternity. They were aware that, twenty-six centuries earlier, the Assyrian king Esarhaddon declared "I had monuments made of bronze and inscriptions of baked clay. I left them in the foundations for future times." The ships bore their own hieroglyphics in that ancient tradition. As with "The Inner Light," they are inscribed with a message created to be read by the beings of other worlds and times.

Like the Kataani, human scholars wondered what we humans could have in common with sentient beings with their own evolutionary history. Perhaps a history so advanced that they are able to sail galactic space. The answer? Math and science. After all, an alien civilization can only sail the galactic sea once they've mastered the mathematical physics of spaceflight, so the ships' message was first based on hydrogen, the most common element in the Universe. The electron in a hydrogen atom flips at a constant rate, so hydrogen atoms are natural clocks, a time-keeping device that we have in common with alien beings.

The next part of the message fixed our cosmic address in space and time. As pulsars are rapidly rotating neutron stars that give off periodic pulses of radio waves, they can act as a time device, so the ships' message showed our Sun at the center of a diagram from which radial lines emanated to the fourteen closest pulsars. A simple cypher shows each pulsar to have a unique frequency, using the ticking of the hydrogen electron as its unit of time, so extraterrestrials could interpret the ships' diagram to find its home star in the Galaxy.

Those two human ships were Voyager 1 and 2. Spacecraft with a projected shelf life of one billion years. Imagine yourself as an alien scientist. You've found a Voyager Golden Record, on a lone ship fished out of the dark illimitable ocean. The ships' message was created one thousand million years ago. What on Earth would you make of the message, of its senders and their ancient world?

To help your alien imagination along, consider the contents of the Voyager Golden Records. They were selected for NASA by a committee chaired by Carl Sagan. They assembled 115 images and a variety of natural sounds that includes a Saturn V rocket launch, those sounds made by surf, wind, and thunder, the songs of birds and whales, and other animals. To this they added the music and greetings in 59 human languages, from different cultures and eras, and the brain waves of a young woman newly fallen in love. And the sound of a pulsar. All of which will live for a billion years.

CITY ON THE EDGE OF FOREVER: THE SCIENCE OF ALLO-HISTORIES

"For most of the twentieth century, modern cultural accounts of alternate history have been associated with science fiction. Typically, such histories blend with time travel. Jumping from one history to its alternate, and awareness of the presence of one timeline by the people in another, is a common theme of the genre. Indeed, cross-time alternate histories have become so closely related that it's almost impossible to separate them from the genre as a whole."
—Mark Brake, *The Science of Science Fiction* (2018)

"Sometimes it's possible, just barely possible, to imagine a version of this world different from the existing one, a world in which there is true justice, heroic honesty, a clear perception possessed by each individual about how to treat all the others. Sometimes I swear I could see it, glittering in the pavement, glowing between the words in a stranger's sentence, a green, impossible vision—the world as it was meant to be, like a mist around the world as it is."
—Ben H. Winters, *Underground Airlines* (2016)

"Science fiction writers write it. And it uses a very science fictional technique: change one thing and extrapolate from that."
—Harry Turtledove, on alternate history, interview at Worldcon (2001)

CITY ON THE EDGE OF FOREVER

"The City on the Edge of Forever" was the twenty-eighth and penultimate episode of *The Original Series*. The story had several writers contribute to the finished product, but most notably Gene Roddenberry's positive humanist vision was combined with the more combative worldview of American science fiction writer Harlan Ellison. Ellison was once described by Robert Bloch, the author of *Psycho*, as "the only living organism I know whose natural habitat is hot water." (Bloch was clearly unaware of extremophiles.)

"The City on the Edge of Forever" sees Kirk and Spock follow McCoy to the surface of an unknown planet. McCoy has accidentally injected himself, as doctors do, with a drug that causes paranoid delusions. Down on the planet the crew find a donut-shaped sentient formation known as the Guardian of Forever. (Sentient entities often have such pretentiously grandiose names in science fiction.) This particular entity can transport people to any place in spacetime, functioning as a portal to a time vortex that allows access to other times, locations, and dimensions. Like a kind of static, funfair TARDIS. Very handy.

But the delusional McCoy leaps through the portal without warning and is transported back in time. Kirk and Spock quickly find that whatever McCoy has *just* done in the past has major repercussions, which include somehow causing the Enterprise to un-exist! To correct this more than minor inconvenience in their time-stream, Kirk and Spock travel back to stop McCoy's mysterious actions, which kick-starts the main narrative of "The City on the Edge of Forever."

MANY WORLDS, MANY TIMELINES

Before we consider "The City on the Edge of Forever" in more detail, let's look at some background science and philosophy. One of the main philosophical questions asked about time and history is this: Do we live in the best of all possible worlds? One would certainly hope the answer is no, given the future possibility of prospects such as nuclear war, bio-engineered pandemics, or malign artificial superintelligent software, to name but three.

This question about living in the best possible world was first raised by German philosopher Gottfried Leibniz. Leibniz was a brilliant polymath who not only invented a branch of math called calculus, independently of Isaac Newton, but also refined the binary number system, which today is the basis of almost all digital computers. Clearly not a man to shirk the tricky questions, Leibniz turned his keen intelligence to try solving the problem of evil. His thinking went something like this: If God is good, omnipotent, and omniscient, how come there is so much suffering and injustice in the world? Leibniz's solution in many ways preempted a science fictional obsession that was to follow, typified by the plot of "The City on the Edge of Forever." Leibniz made God a kind of "optimizer." God simply chose from a host of all original possibilities. And since God is good, naturally, this world must be the best of all possible worlds. Convinced? Neither was atheist Harlan Ellison, but more of that later.

Leibniz's question found an echo in the science of quantum mechanics. Chief among the mysteries of quantum theory is the notion that our Universe is merely one of many. This is known as the "many-worlds interpretation." It was promoted by the likes of American physicist John Wheeler and popularized by science fiction writers the world over. But the theory goes further. Some proponents imagine an infinite number of parallel worlds or Universes, making up a "multiverse" that together comprise all of physical reality.

If you recall *Star Trek*'s Mirror Universe, a parallel Universe in which several episodes take place, you will be familiar with all this weird stuff about a multiverse in which exist all possible Earthly histories and all possible physical Universes. And if your head hurts, don't worry; it's quite natural. Quantum supremo John Wheeler once said, "If you are not completely confused by quantum mechanics, you do not understand it." Perhaps the words of Philip Pullman in his 1995 novel, *Northern Lights*, will help: "If a coin comes down heads, that means that the possibility of its coming down tails has collapsed. Until that moment the two possibilities were equal. But on another world, it does come down tails. And when that happens, the two worlds split apart."

Star Trek plays with quantum theory in many different forms. Parallel worlds may also be conjured in stories under names such as "other dimensions," "alternate Universes," "quantum Universes," or even "alternate realities." And yet the speculative idea that other worlds lie parallel to ours is older in fiction than it is in the "fact" of quantum theory.

BACK TO THE CITY

"The City on the Edge of Forever" may have been the penultimate episode of the first season of *The Original Series*, but its development began before the first episode of *Star Trek* even premiered. Gene Roddenberry hired Harlan Ellison as one of the first writers for the show in the hope that his talents would help with the series' success. Ellison had been nominated by the Writers Guild of America for his *Outer Limits* script, "Demon with a Glass Hand," so he was given reasonable artistic freedom to develop his own episode. But Ellison's draft of "The City on the Edge of Forever" has many notable differences from what eventually aired.

Back in the day, *Star Trek* was without a series bible that would direct the overall philosophy for each story's character and tone, so Starfleet was a military agency. The traveler to the past was not McCoy but a corrupt lieutenant transported to the planet for execution. The planet itself was home to an ancient, ruined city with huge aliens, and the Enterprise host to a bunch of renegades after time alterations, though reports of the minutiae of the contents of the original outline differ.

NO BIG DONUT

Like the changeable version of time depicted in the tale, the script itself was also subject to much change and revision. But Ellison's main story threads, characters, and themes were kept in. Years later, Ellison said that the inspiration for his celebrated story "was an image of two cities which is what it says in the script 'The City on the Edge of Forever,' [including] the city on this planet. It was not a big donut in my script; it was a city. That city was on the edge of time, and it was where all the winds of time met. When you go through to the other side, here is this other city which is also on the edge of forever, which is New York City during the Depression."

Ellison said that the two cities, New York on Earth and the alien city, were mirror images of one another. Furthermore, "At the time, all I was concerned about was telling a love story . . . I made the point that there are some loves that are so great that you would sacrifice your ship, your crew, your friends, your mother, all of time, everything in defense of this great love. That's what the story was all about." (Incidentally, in Ellison's 1996 book, *The City On the Edge of Forever*, he wrote that he got no more than a mere "pittance" for his work on the episode: "every thug and studio putz and semi-literate bandwagon-jumper and merchandiser has grown fat as a maggot in a corpse off what I created," citing this as the reason he wrote his book, to correct the record.)

MEET EDITH KEELER

But readers might recall that quite a different tale is told in the episode when it actually aired. Kirk and Spock pass through the Guardian of Forever "donut" and find themselves in New York City in the year 1930—a time that precedes McCoy's wrecking intervention. Kirk and Spock try to blend in (not an easy ask for Spock!) as they plan to prevent the McCoy catastrophe. They meet Edith Keeler, a good woman who runs a 21st Street Mission. Naturally, Kirk falls in love with her (one often wonders if the series bible also included a note to the effect that Kirk would forever be susceptible to the ways of women).

Meanwhile, Spock studies his tricorder data to find the truth behind McCoy's time-splicing antics. Keeping an element of tragedy from Ellison's version of the time-tale, Spock finally finds out that Keeler must die so that the course of time may be reset. Were Keeler to live, due to McCoy saving her life from an accident, her pacifist campaigning would keep the United States out of World War II for so long that Nazi Germany would conquer the world. (A hypothetical Axis victory in WWII is a common idea of alternate history science fiction. Even before "The City on the Edge of Forever" had been written, Philip K. Dick's *The Man in the High Castle*, arguably the most celebrated story that uses Axis supremacy as a dramatic device, had already been published.)

THE PHILOSOPHY OF *STAR TREK*

Given that *Star Trek* is a franchise which usually values peace over war, this episode stands out. The story is clearly saying that pacifism can lead to ruinous results. As with many time-travel tales, in "The City on the

Edge of Forever," the course of human history hangs on the apparent quantum mechanical minutiae of life: weighing personal well-being against the greater good. This episode is an early example of an idea that would help characterize many future *Star Trek* stories: "the needs of the many outweigh the needs of the few," a refrain we hear in "The Wrath of Khan" and "The Voyage Home," among others.

Warp drives and blustering space battles notwithstanding, much of the philosophy of *Star Trek* is derived from the actions of individuals making difficult decisions in the face of ethical and societal dilemmas. In a March 2008 edition of the BBC Radio 4 program *In Our Time*, Mary Beard, Professor of Classics at the University of Cambridge, spoke about the Greek myths as a kind of thought experiment. She said that the myths were a kind of moral and philosophical reference point in the absence of anything else that had the same kind of cultural dissemination. In short, the Greek myths were "what if?" thought experiments.

In this sense, science fiction like *Star Trek* is a form of modern myth, a modern form of a "what if?" thought experiment. *Star Trek*'s cultural dissemination and influence has always been mightily impressive. And its stories are thought experiments that explore what it is to be human in a Universe that is decentralized, inhuman, infinite, and alien. The crew of starships such as Enterprise face strange new species, new worlds, and artificial life as they roam the fictional Galaxy. And by imagining these strange worlds, we come to see our own conditions of life in a new perspective.

Star Trek is concerned with the nonhuman world revealed by science. And the space-based stories do the job far better than would a drama series in a more grounded setting. Believe it or not, no one described science fiction better than the old English Romantic poet William Wordsworth. In speaking about the poet's interest in science, Wordsworth wrote "If the labors of men of science should ever create any material revolution . . . in our condition . . . the poet will sleep then no more than at present, but he will be ready to follow the steps of the man of science, not only in those general indirect effects, but he will be at his side, carrying sensation into the midst of the objects of the science itself."

Star Trek does what Wordsworth suggested. The series tries to best express "the taste, the feel, the human meaning of scientific discoveries." The stories are a provocative and compelling way to think about the changes brought by science and progress. And the best science fiction of *Star Trek* gives us a way of reducing the gap between the new worlds uncovered by science, and the fantastic, strange worlds of the imagination.

I'M IN LOVE WITH EDITH KEELER

Edith Keeler must die. Not only is this Spock's conclusion, but it also crystallizes the sheer beauty of science fiction. The fate of Keeler puts a personal perspective on the process of causal history. It enables the writers to explore history, and demand of viewers that they consider viewpoints they may otherwise have not done.

Star Trek is full of stories about aliens and time travel, spaceships and cyborgs, androids and the end of worlds. The stories have this in common: they're about the way science may affect our lives in the future. So, it should be no surprise that *Star Trek* is interested in the decisions that lead up to a course of action, and the consequences of such decisions. This is at the very heart of Harlan Ellison's original story.

While the fabric of time and Earth history is at stake, "The City on the Edge of Forever" focuses on Kirk and Keeler's love, leading to the story's tragic conclusion. They grow more intimate and find surprisingly that they share worldviews. Thus, the tragedy grows deeper still. Kirk will soon not only have to sacrifice a loved one, but also lose a kindred futurist spirit. Despite living in the 1930s, Keeler has a pioneering philosophy straight out of a *Star Trek* future: "One day soon, man is going to be able to harness incredible energies, maybe even the atom . . . energies that could ultimately hurl us to other worlds in . . . in some sort of spaceship. And the men that reach out into space will be able to find ways to feed the hungry millions of the world and cure their diseases. They will be able to find a way to give each other hope and a common future. And those are the days worth living for." Little wonder Kirk fell for Keeler, irrespective of what it might or might not say in the series bible!

And so here's the quantum mechanical contradiction in history and time. Keeler must not live to see the future she envisions, for her continued existence is likely to stop such a world from materializing in the first place. "The City on the Edge of Forever" was not the first *Star Trek* episode to center on time travel, nor was it the last, of course. And yet, it is the influence of Harlan Ellison's sharp focus on the personal toll of sacrifice for the greater good that makes this episode so memorable and so influential on many time-travel tales that came later. "The City on the Edge of Forever" remains a hallmark of the poignant humanist philosophy of *Star Trek*.

WHAT'S A HISTORY OF
STAR TREK IN SEVEN OBJECTS?

"We travel back in time and across the globe, to see how we humans have shaped our world and been shaped by it over the past two million years. The book tries to tell a history of the world . . . by deciphering the messages which objects communicate across time . . . They speak of whole societies and complex processes rather than individual events, and tell of the world for which they were made, as well as of the later periods which reshaped and relocated them . . . It is the things humanity has made . . . and their often-curious journeys across centuries . . . which *A History of the World in 100 Objects* tries to bring to life."
—Neil MacGregor, *A History of the World in 100 Objects* (2010)

Jean-Luc Picard: "Live now; make now always the most precious time. Now will never come again."
—Morgan Gendel and Peter Allan Fields,
teleplay of *Inner Light* (1992)

SIGNALS FROM THE PAST

How do you grapple with history? In 2010, the British Museum and the BBC had a bright idea. They presented, over a period of six months on BBC Radio 4, a catalog of one hundred objects of art and technology from the British Museum, which tell a history of the world. The one hundred objects ranged from a Tanzanian million-year-old hand axe, through the statue of a Minoan Bull-leaper from the ancient culture on Crete, to the ship's chronometer from Darwin's HMS *Beagle*. The Museum was clear that this landmark project told "a" history, not "the" history.

And what better way to give a sense of time about the influence of *Star Trek* on modern culture than to compose our own catalog listing. For want of space, I have reduced the list from one hundred objects down to a magnificent seven, in no particular order. But I retain the same intention. To briefly tell the tale of each "object" from my own chosen history of *Star Trek*. If you don't agree with my listing, why not compile your own?

1: KIRK AND UHURA'S KISS

An object of desire. Kirk and Uhura's kiss occurred in *The Original Series* episode "Plato's Stepchildren." The kiss was first broadcast on November 22, 1968, and is often cited as being the first scripted interracial kiss on US television. (Several earlier instances have since come to light, including an unscripted kiss on the cheek between Sammy Davis, Jr. and Nancy Sinatra on *Movin' with Nancy* in 1967 and the very first interracial kiss on British television on February 1, 1959, in the ITV *Armchair Theatre* adaptation of the play *Hot Summer Night*.)

The important thing about Kirk and Uhura's kiss was context. Only the United States had seen a decades-long civil rights movement, in which African Americans and their like-minded allies had campaigned through the use of nonviolent resistance and civil disobedience to bring to an end the country's institutionalized racial discrimination, disenfranchisement, and racial segregation. In April of 1968, Martin Luther King Jr. had been murdered. And prior to the broadcast of the kiss, the civil rights movement had achieved its largest legislative gains in the mid-1960s, after many years of direct action and grassroots protests.

Indeed, Kirk and Uhura's kiss had its own element of direct action and grassroots protest. Here's what happened. At one stage during episode development, it was suggested that Spock kiss Uhura, which was apparently potentially less offensive to the weak-minded, as Spock was half Vulcan. But William Shatner demanded they stick with the original script. As a compromise, NBC said that two versions of the scene should be shot: one in which Kirk and Uhura kissed, and one in which they didn't. With the kiss version of the scene successfully in the can, Bill Shatner and

Nichelle Nichols purposely botched every take of the non-kiss version, so pushing the episode through to its kiss conclusion.

As Nichelle Nichols wrote in her 1994 book, *Beyond Uhura: Star Trek and Other Memories*, "Gene was determined to air the real kiss . . . The director was beside himself, and still determined to get the kiss-less shot . . . The next day they screened the dailies . . . When the non-kissing scene came on, everyone in the room cracked up. The last shot, which looked okay on the set, actually had Bill wildly crossing his eyes. It was so corny and just plain bad it was unusable." Nichols said it became clear that the only alternative was to cut out the scene altogether. But that was impossible to do without ruining the entire episode. She concluded, "Finally, the guys in charge relented: 'To hell with it. Let's go with the kiss.' I guess they figured we were going to be cancelled in a few months anyway. And so the kiss stayed."

Whoopi Goldberg, who later played Guinan in *Next Generation*, recalled that the impact of seeing Nichols's Uhura was life changing. "[W]hen I was nine years old, *Star Trek* came on. I looked at it and I went screaming through the house, 'Come here, mom, everybody, come quick, come quick, there's a black lady on television, and she ain't no maid!'"

Indeed, Martin Luther King Jr. himself considered Nichols's Uhura to be "the first non-stereotypical role portrayed by a black woman in television history." When Nichols was reported as thinking of quitting the show for Broadway, King convinced her to stay. She recounted in *Beyond Uhura*, "Dr. Martin Luther King, quite some time after I'd first met him, approached me and said something along the lines of 'Nichelle, whether you like it or not, you have become a symbol. If you leave, they can replace you with a blonde-haired white girl, and it will be like you were never there. What you've accomplished, for all of us, will only be real if you stay.'" Nichols saw that her role, and indeed the kiss, was bigger than just her.

2: UNIVERSAL HEALTH CARE

An object of a civilized society. In the *Star Trek* franchise, the likes of Captains Kirk and Picard never speak of medical insurance woes. And that's because it's just not necessary in this future landscape of free universal health care. No matter whether it's human, alien, or Borg that needs treatment, this future health-care system treats all alike. (Indeed, one might speculate that a health-care system focused on profiting from the care of the sick would appeal only to the Ferengi.)

Star Trek was beamed out as a progressive broadcast at the height of the US civil rights movement and the Cold War. It portrayed a superior, advanced human society, one in which white Americans peacefully lived and worked side by side on exploratory missions with not only aliens, but also with Russians such as Chekov, people of Asian descent like Sulu, as well as African Americans such as Uhura. This mattered in the larger cultural context.

The free universal health-care system is all part of Gene Roddenberry's original philosophy. Roddenberry explained that the show's creators resisted the idea that television audiences were too dumb or reactionary to understand the show's philosophy: "We believed that the often ridiculed mass audience is sick of this world's petty nationalism and all its old ways and old hatreds, and that people are not only willing but anxious to think beyond those petty beliefs that have for so long kept mankind divided." Roddenberry recognized the intelligence of his audience: "So you see that the formula, the magic ingredient that many people keep seeking and many of them keep missing is really not in *Star Trek*. It is in the audience. There is an intelligent life form out on the other side of that television, too . . . "

3: GUARDIAN OF FOREVER

An object of temporal destruction. Sure, we might have just spoken about the Guardian of Forever in the previous chapter. And true, it might be risibly donut-shaped and suffer from a pretentiously grandiose name. But what other time portal is not only powered by sentient alien intelligence, but can also allow travel to anywhere in time or space?

The seeds of destruction are often present in time-travel tales. One only has to think of Marty McFly in the 1985 movie *Back to the Future*. The more Marty messes with his parents' past, the more he begins to feel himself fading away. But this potential for destruction seems to be writ large with the Guardian of Forever. There is the eternal possibility that someone going through its torus could destroy their own future.

We know this happens in "The City on the Edge of Forever." A time-transplanted McCoy first saves Edith Keeler's life, allowing her political activism to delay US intervention in World War II, which then leads to Nazi domination of planet Earth. As time distortions go, that's pretty destructive. That's just the way the Guardian rolls. Whether by design or not, the Guardian threatens to unravel the very thread of reality every time someone jumps through its torus. And that, along with its vast sentient intelligence that spans all of spacetime, makes the Guardian of Forever one of the franchise's most powerful artifacts.

(A Guardian of Forever reprise was planned for *The Next Generation*. The plot was proposed by script writers Trent Christopher Ganino and Eric A. Stillwell during the creation period for the episode "Yesterday's Enterprise." Their plot envisaged a team of Vulcan scientists, led by Sarek, sent to study the Guardian. But the team are accidentally sent back in time to the past on their planet, where they alter history by causing the death of Surak, the founder of Vulcan logic. On their return, the team find their people in alliance with the Romulans. And at war with the Federation. So Sarek returns to the planet's past and takes on the mantle of Surak to restore the timeline. But producer Michael Piller wanted the episode's plot to be focused on the main cast of *Next Generation*, so Sarek and the Guardian were ditched from the story.)

4: DYSON SPHERE

An object of an advanced extraterrestrial civilization. We speak elsewhere in this book about the *Next Generation* episode "Relics," in which the Enterprise responds to a nearby distress call and discovers a Dyson Sphere. The Sphere is an enormous hollow structure, built around a star, which would be able to harness all the star's radiant energy. A civilization living on the interior surface of the Sphere would have astronomical sources of power.

But let's think again about the name of this Dyson Sphere episode. "Relics." We might define a relic as an object, tradition, or system from the past that continues to exist. And what a relic the Dyson Sphere is. In 1964, Soviet astrophysicist Nikolai Kardashev proposed three levels of civilizations in space. And these levels were based on the order of magnitude of power available to them. Kardashev's scale has three designated categories. Type I civilizations, also called planetary civilizations, can use and store all of the energy available on its planet. Type II civilizations, also known as stellar civilizations, can use and control energy at the scale of its stellar system. And Type III civilizations, also called galactic civilizations, can control energy at the scale of its entire host Galaxy. We humans have not yet reached Type 1 civilization status. But the Type II civilization that built the Dyson Sphere was clearly capable of harnessing the energy radiated by their own star.

Apart from the fascinating speculative future science, the Dyson Sphere has a place in our Seven Objects for the following reasons. The franchise is frequently credited with predicting future tech. You know the kind of thing propeller-heads get excited about—a world of cell phones and Facetime calls decades before the actual tech became possible. But "Relics" shows that *Star Trek* also helps us visualize pre-existing futuristic concepts from science itself. And the Sphere is arguably one of the most amazing engineering feats ever seen by the franchise's characters.

5: KATAAN PROBE

An object of a bygone civilization. The Kataan Probe appeared in the *Next Generation* episode "Inner Light." The probe was an artifact created by the Kataan, a humanoid race on a planet that seems to be on the cusp of post-industrial revolution. When the people realized their own demise was at hand, due to their Sun going slowly nova, they launched the probe, which held a poignant memory record of their civilization. The probe, of course, ultimately found Picard and decanted its memory record into him, gifting him their life experiences in a matter of minutes.

As it turned out, this habit of ancient alien cultures imposing their past on people is an experience the *Next Generation* crew suffer more than once. What's exceptional about the Kataan Probe is the big punch it packs. Picard was never the same again. The relic reminds us of how the lifetimes of such probes in the future could be far longer than the lifetime of the civilization that built it. As with the Kataan Probe, the relic dwells in interstellar space for millions of years after its parent civilization had passed. The use of the probe to transmit the heritage treasures of the culture of the dead Kataan civilization into the cosmos for hundreds of millions of years is truly ingenious.

6: PRESERVER OBELISK

An object of true antiquity. Obelisks have a prominent history in science fiction. Arguably the most famous obelisks appear in Arthur C. Clarke's Space *Odyssey*. Here, the obelisks are monolithic machines built by an unseen alien race. In the series of Clarke's novels and associated movies, the obelisks encourage humankind to progress with tech evolution and space travel.

In *The Original Series* episode "The Paradise Syndrome," we meet another obelisk. This too is an artifact built by an elusive alien race and is a very powerful type of tech. Easily underestimated, as it looked like a 1960s piece of statement art, the obelisk proceeds to zap Kirk's memories, leaving him to fend for himself on an alien world for many weeks. Once Kirk's memory is restored by Spock, the crew members learn that the obelisk was created by the Preservers to help Native Americans ward off asteroids with more power than Enterprise could provide. But the obelisks had an ancient backstory. Constructed from a "compound" which only existed at the hearts of white dwarf stars, 119 obelisks had been discovered by the year 2370. They ranged in age from a mere six years to over two billion years old.

7: TIME CRYSTAL

An object of coronal displacement. *Star Trek* has, naturally, had its fair share of time-travel artifacts. And yet galactic con-man Harcourt Fenton Mudd introduces a new type of time-travel tech in the form of the time crystal, which enables the wielder to do basically anything they want. Mudd torments the crew of the Discovery with a type of time crystal which enables iterations of the same time loop (while Mudd still retains his memories). This allows a kind of cosmic Groundhog Day, which lets Mudd have as many as he needs for the crime of his choice. And, of course, it also means he cumulatively learns more about his enemies. Unfortunately, the whole system went to pot once said loop was broken.

Curiously, the ancient Greeks had a similar circular notion of time. Philosophers such as Plato, Aristotle, and even Pythagoras believed that time was cyclical, and that the history of the cosmos was made up of a series of "great years," as they called them. Each cycle of unnamed length ended in a planetary conjunction, which unleashed an apocalypse. Then, a new cycle began out of the ashes of the old.

Unlike the *Star Trek* time crystal, though, this Greek notion of time left little room for the idea of evolution. Instead, the past was as closed and confining as the cosmos in which Aristotle kept space captive. It was part and parcel of a general feeling that history was eternal. There was no real beginning in time, since the "great years" ran endlessly on. Or, as Eudemus of Rhodes put it to his pupils, "If you believe the Pythagoreans, everything will eventually return in the self-same numerical order, and I shall converse with you staff in hand, and you will sit as you are sitting now, and so it will be in everything else." A sobering thought!

VOYAGE HOME TO SAVE THE WHALE: WHICH EXTINCTIONS MOST CHANGED TIME?

"The key to saving the future, can be found only in the past."
—*Star Trek IV: The Voyage Home* tagline (1986)

"Jonbar Point: term used for a crucial forking-place in time, whose manipulation can radically affect the future that follows. The name derives from Jack Williamson's *The Legion of Time* (1938), which deals with the potential future empires of Jonbar (good) and Gyronchi (bad). The former is named for the character John Barr: the fiercely contested jonbar point is the moment when as a small boy Barr picks up either a magnet, inspiring him to a life of science which ultimately brings Jonbar into existence, or a pebble, leading Barr to obscurity and the world to Gyronchi."
—John Clute and Peter Nicholls,
The Encyclopedia of Science Fiction (1979)

SPACESHIP THEFT

Science fiction is replete with aliens stealing spaceships. In *Doctor Who*, the Doctor steals the TARDIS. (In the story "The Doctor's Wife," the soul of the TARDIS is decanted into the body of a humanoid female called Idris, who tells the Doctor she allowed the Doctor to "steal" her because she wanted to see the cosmos for herself.) In *The Hitchhiker's Guide to the Galaxy*, Zaphod Beeblebrox, the President of the Galaxy, steals the Heart of Gold, a spaceship that uses a unique Infinite Improbability Drive, which handily allows it to travel instantaneously to any point in spacetime. And in *Star Wars: The Force Awakens,* Scavenger and stormtrooper Finn steal the Millennium Falcon to escape an attack by the First Order.

But in *Star Trek IV: The Voyage Home,* the plot is far thicker. The motley former crew of the USS *Enterprise,* living in exile on the planet Vulcan, use the pirated starship from *Star Trek III,* after receiving a distress call from Earth when they're already returning to their home planet to hear their court martial. It seems that an enormous cylindrical space probe has entered Earth's orbit, disabled global power, generated planetary storms, and created catastrophic, Sun-blocking cloud cover. Tricky. So, Kirk, Spock, and the rest of the officers travel back in time to rescue some now-extinct humpback whales. Why? Because Spock has deduced the wise old whales can communicate with the probe and send it packing away from our planet.

WHAT-IFS

The plot of *The Voyage Home* reminds us of that popular science fiction question of "what if?" What if the humpback whales had not become extinct? Trek's motley crew need not have used their stolen spaceship. Kirk and co need not have traveled back in time. And so on. Science fiction is full of such "what if?" questions. What if history had happened so very differently? What if grass had died? What if Neanderthals had not become extinct? What if Earth had a habitable Moon instead of a dead lunar satellite, or a series of alien invasions had made the world well aware of life on other planets before the Renaissance? What if Christopher Columbus never sailed west? What if the British Empire had conquered China? What if there was no oil in the Middle East? What if the South had won the US Civil War? What if Hitler had successfully invaded Russia? What tiny amount of tinkering in time would result in the most profoundly different present?

British novelist L. P. Hartley once said, "The past is a foreign country; they do things differently there." But in alternate history stories, things in the past did happen differently, and the present becomes a foreign country, of sorts. Alternate history stories contain "what if" scenarios, like those listed above, which revolve around crucial "Jonbar" points. A Jonbar point, or Jonbar hinge, is the idea of a vital point of divergence between two outcomes, especially in time-travel stories. Sometimes referred to as a change point, this idea helps focus on those points in the past that could have had a different outcome to that recorded in actual history.

NATURAL SELECTION AND LIFE'S PATHWAY

Nature has its own "Jonbar" points. Crucial points of divergence where life's pathway could have had a markedly different outcome. Outcomes that are very unlikely to have led to human consciousness. Our theory of evolutionary change helps explain and understand the actual pathway of life's history. How we got where we are today. That evolutionary theory first involves natural selection. Natural selection locates the mechanism of evolutionary change in the struggle among organisms for reproductive success. And this leads to an improved fit of a population to a changing environment. But, as Darwin himself said, "Natural Selection has been the most important, but not the exclusive, means of modification."

Natural selection doesn't paint the whole picture of evolutionary change. There are two good reasons for this. First, there are other crucial causes of change. They occur at levels both below and above the conventional Darwinian focus on that adaptive struggle of organisms for reproductive success. Below the traditional focus, we have DNA, the neutral and random change associated with the mutations in individual base pairs of DNA. Above the traditional focus, we get mass extinctions. These can involve entire species. They can wipe out substantial parts of biotas for reasons unrelated to adaptive struggles of constituent species in the times between such events.

Second, and our concern in this chapter, is that irrespective of our general theory of evolution, we also need to understand the actual pathway of life's history on Earth. Humans didn't just appear on Earth, just a geologic second ago, because theory predicts so, based on ideas of progress and increasing complexity. No, homo sapiens arrived instead via a lucky contingent set of outcomes, thousands of linked Jonbar events. And any one of those Jonbar points could have occurred differently. Any change in the point could have sent history on an alternative pathway that would not have led to human consciousness.

WHICH EXTINCTIONS MOST CHANGED TIME?

Which natural and evolutionary Jonbar points most changed time? This somewhat science fictional question helps us understand the actual pathway that life eventually took. Our theories are useful in predicting some general aspects of life's pathway, but the actual pathway is not slavish to our theories. This point is vital in understanding our planet's complexity. The vast network of historical events is incredibly intricate, infused with myriad millions of chaotic and random factors. And those factors influence such a multitude of unique objects that our models of prediction won't work.

Here, hindsight is everything. History may be explained after a series of events has occurred. But it cannot be precisely predicted beforehand. Science is not prophecy. Our past contains too much chaos, too much dependence on tiny and unmeasurable variations in initial conditions. And such nuance leads to hugely divergent outcomes, which are based on those minute and unknowable differences in starting points.

In short, history has too much contingency. Too much potential in so many Jonbar events. Our present is shaped by huge networks of unpredictable precursor points in the past. And human consciousness was not determined by timeless laws of nature. So, which extinctions most changed time? Let's cherry-pick some standouts from nature's host of possibilities.

THE CAMBRIAN EXPLOSION

Jonbar one: the Cambrian explosion. What if our modest and fragile lineage hadn't been one of the few survivors of the Cambrian explosion 541 million years ago? The Cambrian radiation was an event in the Cambrian geological period, which lasted for 13–25 million years. It was a period when almost all major animal phyla began showing up in the fossil record. A phylum is a group of organisms that have the same body plan. If our group had not been among the initial radiation of multicellular animal life, then no vertebrates would have inhabited the Earth at all.

(A single member of our chordate phylum, the genus Pikaia, has been found among these earliest fossils. Resembling the lancelet, and perhaps swimming much like an eel, Pikaia has gotten lots of attention among the multitude of animal fossils found in the famous Burgess Shale in the mountains of British Columbia, Canada. For example, in 2012, researchers from the University of Cambridge, University of Toronto, and the Royal Ontario Museum confirmed that a 505-million-year-old creature, found only in the Burgess Shale fossil beds in Canada's Yoho National Park, was the most primitive known vertebrate and thus the ancestor of all descendant vertebrates, including humans.)

THE LOBE-FINNED FISHES

Jonbar two: the lobe-finned fishes. What if that small and unpromising group, the lobe-finned fishes, had not evolved? The lobe-finned fishes diverged from the ray-finned fishes about 450 million years ago. They were very successful during the Devonian Period, when hundreds of species lived in the oceans and rivers of most continents. They had fin bones with a strong central axis capable of bearing weight on land. More than 350 million years ago, our distant fishy ancestors traded in the life aquatic for life on land. Once ashore, the four-limbed vertebrates, known as tetrapods, branched into an impressive range of animals: amphibians, reptiles, dinosaurs, birds, and mammals.

PREHISTORIC FISH EXTINCTION

Jonbar three: the mass extinction of fish. We also owe our place on the planet to a mass extinction of fish 360 million years ago, according to recent research. This cataclysmic Jonbar point re-spawned the evolutionary growth point for all vertebrates living today. If this Jonbar point had gone another way, human consciousness either may not have evolved, or could have evolved quite differently. Crucial traits shared by modern mammals, such as five-digit limbs, began when life was reset after this mass extinction.

The extinction, which occurred in the Devonian Period between 416 and 359 million years ago and is sometimes known as the Age of Fishes, was global. A broad array of species had filled the oceans, lakes, and rivers, but most were unlike any alive today. Armored placoderms, like the monstrous 30-foot carnivore Dunkeosteus, dominated the waters. Ray-finned fishes, sharks, and four-limbed tetrapods were in the minority. But the picture changed abruptly and dramatically with the traumatic Jonbar event known as the Hangenberg extinction. It kickstarted vertebrate diversity in every single environment, both freshwater and marine, and birthed an entirely new world. The slate of life was very nearly wiped clean and, of the few stragglers that made it through, a handful then re-radiated spectacularly. What happened to trigger the mass extinction remains an unsolved mystery. But the Hangenberg event reset life's pathway and led eventually to us humans. (Assuming, dear reader, that you are human.)

COMET CRASH

Jonbar four: comet crash and dinosaur demise. One of the biggest impact craters on planet Earth, at over 100 miles wide and 3,000 feet deep and dubbed the Chicxulub crater, is buried underneath the Yucatán Peninsula in Mexico. Dated at 66 million years old, and located at the Cretaceous–Paleogene boundary, this smoking-gun crater is one of our planet's largest crater sites, a measure of just how big that fateful day at the end of the Cretaceous really was. Maybe the biggest comet to crash into the Earth over the last half billion years or so. And yet, were it not for this chance collision with a comet, dinosaurs may still be dominant and mammals insignificant, as this situation had prevailed for 100 million years previously. (Some readers may recall an episode of *Star Trek: Voyager*, "Distant Origin," where Voyager encounters an alien race, the Voth, who are humanoid lizards. Captain Janeway, with the help of a scientist called Professor Gegen, used simulations to find that the Voths descended from a species of dinosaur known as the Hadrosaurs, of the genus Parasaurolophus. This story begs so many questions, if the Voth did indeed originate from Earth. For example, how did the Voth get into the Delta quadrant? Did they invent some kind of spacecraft? And were they the first Earth species to invent the warp-drive?

WALKING ON TWO LEGS

Jonbar five: walking upright. What if our lineage of primates had not evolved upright posture? The earliest humans climbed trees and walked on the ground. Such flexibility aided their mobility in diverse habitats and helped cope with changing climates. From between six and three million years ago, our ancestors combined apelike and humanlike ways of getting about. Fossil evidence shows a transition from climbing trees to walking upright on a regular basis. The oldest evidence for walking upright comes from one of the earliest humans, Sahelanthropus. Walking on two legs may have helped Sahelanthropus survive in the diverse habitats of forests and grasslands.

Jonbar five has a volcanic backstory. A sequence of volcanic eruptions had ended the Triassic world. But a disastrous extinction event for one species is a golden opportunity for another. The Triassic extinctions gifted one group the chance to take center stage. The terrible lizards had a 170-million-year run, but when molten rock once more burst through the Earth's surface and flooded a huge area of western India, few animals larger than a hundred pounds survived the catastrophes of the late Cretaceous. The resulting dust cloud brought night to the planet's surface for months. The dinosaurs starved to death. But there were lesser creatures who found some sanctuary in Armageddon. And soon they found that the terrible lizards who had hunted them were gone.

Meanwhile, in one very remote corner of the world, a channel that had separated North and South America, letting sea ocean currents flow from the Atlantic into the Pacific, was slammed shut by tectonic forces, creating the Isthmus of Panama. The event reset the world's network of ocean currents, profoundly affecting the global climate. This meant that in Africa, many lush green forests became sparser landscapes. And species that were experts limited to living life in the trees became extinct.

But the generalists, like our ancestors, found other ways to make a living. We abided and evolved. We had once burrowed deep in the Earth to avoid the predators above. But when the terrible lizards died off, over the ages we made new lives in the trees. We developed opposable thumbs and toes for swinging across the treetops, where all our needs were met. But we could also walk upright, even if just for short distances.

Then, when the climate in Africa got colder, and the trees began to thin out, savannas sprang up. Our ancestors evolved the ability to walk great distances on their hind legs and to run when the big cats came. And this changed the way we looked at the world. With hands no longer needed for walking, they were free to gather food and wield sticks and bones as weapons and tools.

That's the Jonbar five backstory: a change in the topography of a piece of land half a world away reroutes sea currents, Africa grows drier, the trees don't like the new climate, and the primates who dwell in them have to seek new homes. In no time at all, they're wielding tools to reshape the planet.

Every time you take a step, you momentarily stand on one leg, thereby adding stress to your leg bones. The early humans known as Australopithecus anamensis had a wide area of bone just below the knee joint, which is strong evidence that they walked upright. Fossils from around four million years ago, which come from early human species that lived near open areas and dense woods, show that their bones had evolved in such a way as to enable them to walk upright most of the time, but still climb trees. Thus, they could take advantage of both habitats.

Fossils from around two million years ago, of the early human known as Homo erectus, show how similar they are to modern humans. This early human was able to walk long distances. That ability was a great advantage during this time, as East Africa's environments were fluctuating widely between moist and dry, and open grasslands were beginning to spread. As our ancestors became more intelligent, upright posture enabled them to explore more distant and colder places. Over time, this meant that humans could spread out and gradually inhabit most areas of the planet. If it were not for walking upright, our ancestry might have ended in a line of apes that, like chimps and gorillas today, would have become ecologically marginal and maybe doomed to extinction despite their intelligence and complexity.

So, there we have it. *Voyage Home* plays with the idea of the hidden and unexpected dangers that could lurk around the corner, should diversity on Earth diminish. And this fantastic story of the extinction of humpback whales has echoes in real science. Any change in millions of

Jonbar points—the Cambrian explosion, lobe-finned fishes, Hangenberg extinction, comet crash, and hominid evolution included—means that history could have quite easily followed an alternative pathway that may not have led to human consciousness.

WHAT DOES "THE NAKED NOW" TEACH US ABOUT ROCKETS AND STARS?

"We inhabit a Universe where atoms are made in the centers of stars; where each second a thousand Suns are born; where life is sparked by Sunlight and lightning in the airs and waters of youthful planets; where the raw material for biological evolution is sometimes made by the explosion of a star halfway across the Milky Way; where a thing as beautiful as a Galaxy is formed a hundred billion times—a Cosmos of quasars and quarks, snowflakes and fireflies, where there may be black holes and other Universes and extraterrestrial civilizations whose radio messages are at this moment reaching the Earth. How pallid by comparison are the pretensions of superstition and pseudoscience; how important it is for us to pursue and understand science, that characteristically human endeavor."

—Carl Sagan, *Cosmos* (1980)

THE SS TSIOLKOVSKY

In the *Star Trek: The Next Generation* episode "The Naked Now," the crew of the Enterprise receives messages from the SS *Tsiolkovsky*, a science vessel monitoring the collapse of a supergiant star. Before we launch into the science of that collapsing supergiant star, consider the choice of the name Tsiolkovsky, as it's an interesting one. Konstantin Eduardovich Tsiolkovsky was a Soviet rocket scientist and pioneer of astronautics.

Though Tsiolkovsky was a recluse who lived most of his life in a log cabin in Kaluga, well over one hundred miles southwest of Moscow, he was nonetheless the father of astronautics and human spaceflight. No one comes close, save perhaps the Frenchman Robert Esnault-Pelterie, the Transylvanian-German Hermann Oberth, and the American Robert H. Goddard. But even these rocket luminaries follow in the slipstream of Konstantin.

Indeed, the scientific work of Konstantin Tsiolkovsky reads like a real-life Zefram Cochrane. Tsiolkovsky wrote over four hundred treatises, including around ninety published essays on space travel and associated topics. His published interests included space stations, designs for rockets with steering thrusters, multistage boosters, airlocks for exiting a spaceship into the vacuum of space, and closed-cycle biological systems to provide oxygen and food for the first colonies in space.

Tsiolkovsky launched the space race. Witness his words, spoken at the age of seventy-seven on May Day in 1935 from atop Lenin's Mausoleum: "I am finally convinced that a dream of mine, space travel, for which I have given the theoretical foundations, will be realized. I believe that many of you will be witnesses of the first journey beyond the atmosphere. In the Soviet Union we have many young pilots . . . (and) I place my most daring hopes in them. They will help to actualize my discoveries and will prepare the gifted builders of the first space vehicle. Heroes and men of courage will inaugurate the first airways: Earth to Moon orbit, Earth to Mars orbit, and still farther; Moscow to the Moon, Kaluga to Mars!" Less than a generation later, the Soviet Union launched the world's first artificial satellite into space. After *Sputnik-1*'s flight into space on October 4, 1957, the satellite entered into legend for launching the space race. None of this would have happened without Tsiolkovsky, who was not only a brilliant mathematician and engineer, but also a science fiction visionary.

"BESIDES BOOK, I HAD NO OTHER TEACHERS."

To Tsiolkovsky, science fiction was just as important as science. While still less than twenty years old, he moved up to Moscow to study at the world-renowned Rumyantzev Library, known today as the Russian State

Library. In the Rumyantzev he found the works of French science fiction writer Jules Verne. At the time, many of Verne's space tales such as *From Earth to the Moon* and *Off on a Comet* were hugely popular around the globe, and Tsiolkovsky became fascinated by Verne's fantastic accounts of rocket propulsion, space travel, and Moon landings.

(Many Moons before the days of Tsiolkovsky, incidentally, there had lived a man named Wan-Hoo. Wan-Hoo was a minor Chinese official of the Ming Dynasty. He was also the world's first astronaut, or so the story goes. Legend has it that, early in the 1500s, Wan figured he could launch himself into outer space. Cunningly using China's advanced firework technology to his advantage, Wan built his spaceship: a chair. To this chair Wan fastened forty-seven large rockets. Using what influence he had within the Dynasty, Wan conjured up forty-seven assistants. Each willing assistant, armed with a flaming torch, was charged with the task of rushing forward and lighting one of the long fuses. Come the day of liftoff, the finely clothed Wan carefully climbed onto his rocket chair. His forty-seven aides lit the fuses and hastily ran for cover. There was a tremendous roar, a blinding flash of light, and a huge explosion. The smoke cleared. The rocket chair was gone. Wan was never seen again.)

For centuries after Wan-Hoo, science fiction writers had grappled with the idea of propulsion. For example, in his 1865 book, *From the Earth to the Moon*, Jules Verne's preferred method of propulsion was the cannon. Or, rather, the columbiad: a large-caliber, smoothbore, muzzle-loading cannon. The columbiad was capable of firing heavy projectiles at both high and low trajectories. Verne picked a very high trajectory, and an ambitious target: the Moon. (Taking a cue from Verne, Tsiolkovsky created his own science fiction stories. He hoped to spread the science behind his "fantasy." In his 1892 story *On the Moon*, for instance, the main protagonist imagines that he is on the Moon. He describes the alien environment, cold and dark, as well as the weightlessness; the account is vividly factual. For example, when explaining why masses weigh less on the Moon, Tsiolkovsky wrote, "a 30-pound weight is only showing 5 pounds. This means that the force of gravity is reduced by a factor of 6 . . . This exact scale of gravity exists on the surface of the Moon, a result of its smaller volume and the lesser density of its composition."

WHERE NO MAN HAS GONE BEFORE

Tsiolkovsky would later show that a columbiad, like the one used in Verne's tale, would inevitably kill its astronauts due to the extreme force of acceleration. And yet, Tsiolkovsky proposed his own theories of propulsion—theories that wouldn't kill the keen cosmonauts. Tsiolkovsky reasoned that a cosmonaut could leave the Earth if rocket engineers used liquid fuel as the propellant. He worked out the correct ratio of thrust, velocity, and mass in a theory that eventually became known as "Tsiolkovsky's Rocket Equation." His sums were the first scientifically sound suggestion for using rockets to launch into space, and they went on to form the very foundation of contemporary rocket science.

Almost three generations before *The Original Series* aired, Tsiolkovsky wrote about the human colonization of space. In his 1895 work, *Dreams of the Earth and Sky*, before the Wright Brothers made their first flight, Tsiolkovsky was designing space stations, drawing schemas of asteroid mining, and pondering biological growth in space through the use of greenhouses. In his 1932 book, *Cosmic Philosophy*, he wrote about humanity's future traveling among the stars. His philosophy led him to conclude that "the Earth is the cradle of mankind, but one cannot live in the cradle forever." This sentiment was later echoed and quoted by another famous science communicator, American astronomer Carl Sagan.

AS PUMBA SAYS IN *THE LION KING*

Let's rewind to that collapsing supergiant star, the one the SS *Tsiolkovsky* was monitoring in "The Naked Now" episode. What exactly is a supergiant star, and how do they get to explode? Let's start with our local star, the Sun. Would you believe the Sun is a dwarf star? It may be one million miles across, and it may take 227 days to orbit it once, at supersonic speed, but it's still considered small.

Like all stars, the Sun burns gas. As Pumba says in *The Lion King*, "ever wonder what those sparkly dots are up there? I always thought they were balls of gas burning billions of miles away." What Pumba doesn't say is that, in the case of the Sun, a mere 93 million miles away, the temperature at its very center is about 27 million degrees Fahrenheit. And that's hot enough to burn, or fuse, hydrogen gas into helium. So, yes, the Sun is a big ball of fiery energy burning about 4 million tons of gas every second. That's about as much energy as 7 trillion nuclear explosions every second.

And yet, by normal star standards, the Sun is actually quite small. There are many kinds of stars. Some are bright like the Sun, some are dim. The biggest stars are ten million times larger than the smallest ones. Some stars are aged beyond reason, more than ten billion years old. Others are being born as you read this. But the important thing is that the Sun is big enough to burn hydrogen. Hydrogen burning is the main difference between a star, like the Sun, and a big planet, like Jupiter. Stars shine, and planets don't. Planets aren't big enough to get their fires started, even if they're made of star stuff, as Jupiter is. When hydrogen atoms fuse in the hearts of stars, they make starlight.

THE BFG: BETELGEUSE FIERY GIANT

How big do giant stars get? The mass of single stars in an asterism, a pattern of stars that is not a constellation, can range from the runts, not much larger than Jupiter, to the supergiant stars that dwarf the Sun. Let's take a famous example. In the constellation of Orion, there is a red supergiant star by the name of Betelgeuse. (Betelgeuse, also known as Alpha Orionis, is the second-brightest star in Orion, and is the eastern shoulder of the hunter. Its name comes from the Arabic word *bat al-jawzā*, meaning "the giant's shoulder." Betelgeuse is easy enough to spot during stargazing, as it's one of the brightest stars in the northern sky.) If some super deity were to pick up Betelgeuse, and plop it exactly where the Sun sits, at the very center of our solar system, Betelgeuse's belly would be big enough to swallow up the orbits of the four innermost planets: Mercury, Venus, Earth, Mars, and maybe even Jupiter. That's what "The Naked Now" means by a supergiant star. Such monsters have a ravenous appetite for fuel. They burn hydrogen at a furious rate and they have much shorter lifetimes.

Someday soon, cosmically speaking, Betelgeuse will explode in a supernova. The blast will be close enough to blaze brightly during the day, but light-years away so that Earth won't be in peril. Betelgeuse, the first star other than the Sun to be directly imaged, is the nearest red supergiant star to Earth. Far-infrared views of Betelgeuse spied by the

ESA's Herschel Space Observatory reveal multiple arcs, caused by winds from the supergiant crashing against the surrounding interstellar medium, creating a bow shock as the star moves through space.

In September 2019, there was global excitement when Betelgeuse began dimming markedly. This dip in brightness caused some stargazers to predict that a supernova was imminent. But Betelgeuse hasn't gone bang yet. Start stargazing now. Keep an eye on Betelgeuse for yourself. You can't miss it; it's the second-brightest star in Orion, the Hunter's right shoulder. Betelgeuse glows with a subtle orange-red hue, and it's well-situated for stargazing in the first couple months of each year. It's also an ideal star choice for convincing skeptics that stars do, in fact, come in colors.

Betelgeuse will explode. Someday. It's about four and a half thousand light-years from Earth and yet is still one of the brightest stars because it's intrinsically brilliant, around 100,000 times brighter than the Sun. But brilliance comes with a cost. The giant's gargantuan energy means that its fuel burns quickly, hastening its early death. Betelgeuse will eventually run out of gas, collapse under its own gravity, and rebound in a cosmically spectacular supernova. And when Betelgeuse explodes, the giant will brighten hugely over weeks or months, maybe becoming as brilliant as the full Moon, and perhaps even visible in daylight.

Will Betelgeuse's explosion cause destruction on Earth? Probably not. Earth is simply too far away for Betelgeuse's bang to harm life on Earth. Scientists say a planet has to be within fifty light-years of a supernova for it to be dangerous, and Betelgeuse sits four and a half thousand light-years away.

THAT NAKED NOW "RED SUPERGIANT STAR"

Now that we know what a real-life red supergiant star like Betelgeuse does toward the end of its life, let's look again at "The Naked Now." At the start of the episode, Picard reads "Captain's log, Stardate 41209.2. We are running at warp seven to rendezvous with the science vessel SS *Tsiolkovsky*, which has been routinely monitoring the collapse of a red supergiant star into a white dwarf. What has brought us here is a series of strange messages indicating something has gone wrong aboard the research vessel."

This talk of the collapse into a white dwarf appears to happen in a matter of mere minutes. For example, very presently, Worf says, "What we're seeing, sir, is a huge chunk of the star's surface blown away, heading for us." Later, what do we spy heading for the Enterprise? A huge rock! Is this meant to be from the star's surface? We assume so, as Worf then says, "Sir, I estimate fourteen minutes until that mass gets here." This is further confirmed with an entry in the ship's log: "First Officer Riker. Enterprise will be destroyed unless it can be moved out of the path of the star material hurtling toward us." And that suspiciously rock-looking "star material" certainly destroys the SS *Tsiolkovsky*.

And yet, it's not a "red supergiant star" that collapses into a white dwarf, as we have seen with the case of Betelgeuse. Rather, it is more modest stars like the Sun that evolve through a red giant phase and ultimately into a white dwarf. The story should really have gone something like the following.

THE REAL NAKED NOW STORY IS OUR SUN

It's the destiny of stars to collapse. When you stargaze, of the thousands of stars you spy when you look up at the night sky, each lives a life between two collapses: the collapse of birth from a dark, interstellar gas cloud to form a new star, and the collapse of "death" of a luminous star on its way to its fate. It's gravity that makes the new stars collapse and contract. As our Sun is a huge ball of incandescent gas, the reactions at its very core push the Sun to expand outward. But, at the same time, the Sun's gravity field pulls it inward to contract. And so the Sun sits poised in a stable equilibrium between gravity and nuclear explosion, a balance it manages to maintain for billions of years.

Day by day, year on year, millennia after millennia, the Sun consumes hydrogen. And, just as slowly, it shrinks. In response, the Sun's surface gradually expands. But slowly. Imperceptibly. Over the course of millions of years. And in about a thousand million years, the Sun will be about one tenth brighter than it is now. A tenth might not sound like much. And yet that extra Sunlight will have a marked effect on our planet, for when the Sun exhausts its nuclear fire, in five billion years' time, its pressure will drop and its interior will no longer be able to support the weight of its outer layers. The collapse of "death" will kick in.

Even stars die. Or at least they evolve into a different form. Helium ash, the product of ten billion years of hydrogen burning, has amassed at the Sun's core. With no nuclear fusion to hold that weight, there is a collapse of the core until it hits a temperature hot enough to fuse helium into carbon and oxygen. Now, the Sun's core is even hotter than before, and its atmosphere expands. Over the next billion years, the Sun becomes hugely bloated, burgeoning to over one hundred times its ordinary size.

A red giant star is born. This is the kind of star that should have been the subject of *The Next Generation* episode "The Naked Now." A red giant star, not "red supergiant star." It may sound like splitting hairs, but it makes all the difference between the star endpoints of supernova and white dwarf. When the Sun goes red giant, it will expand and devour not only Mercury and Venus but likely Earth too.

Once the Sun has used up its helium, it becomes unstable, sloughing off its outer layers into deep space in the most spectacular way. But not in the way suggested by Worf: a huge "chunk" of the star's surface. Rather, that so-called "chunk" would actually be an expanding, fluorescent shell of gas known as a "planetary nebula," pulsed off a red giant star late in its evolution. (Incidentally, "planetary nebula" is something of a misnomer. Neither planets nor exoplanets, the term "planetary nebula" describes the apparent planet-like shape of these nebulae when first observed by astronomers through the telescopes of the late eighteenth and nineteenth centuries.)

The Sun's naked and super-hot core now floods its neighborhood with high-energy ultraviolet light. Our local star collapses like a soufflé, contracting to the size of the Earth. Further collapse is stopped by its densely overcrowded electrons, which push against any further contraction. The mere seed of light at the center is now the only section of the Sun that abides. A white dwarf star that is but a dim reminder of its past furnace, but one that continues to faintly shine for another hundred billion years.

WHAT DOES THE *STAR TREK: PICARD* ROMULAN LOOK-BACK DEVICE TEACH US ABOUT SCIENCE?

"All creation is a mine, and every man a miner. The whole Earth, and all within it, upon it, and round about it, including himself . . . are the infinitely various 'leads' from which, man, from the first, was to dig out his destiny."

>—Abraham Lincoln, *Discoveries and Inventions* (1860)

"Any one who has studied the history of science knows that almost every great step therein has been made by the 'anticipation of Nature,' that is, by the invention of hypotheses, which, though verifiable, often had very little foundation to start with; and, not infrequently, in spite of a long career of usefulness, turned out to be wholly erroneous in the long run."

>—Thomas Henry Huxley, Collected Essays (1901)

MAPS AND LEGENDS

In the *Star Trek: Picard* episode "Maps and Legends," Picard investigates Dahj's apartment, hoping to find Soji before the assassins get to her. Picard and his Romulan employee Laris beam into Dahj's apartment, where her boyfriend Caler was killed. They find the place lived in, but undisturbed. Laris produces a Romulan forensics device that is used to conduct molecular reconstruction, a noted Romulan science technique. Picard reminds Laris that this technique is illegal in the Federation and that its results are questionable. But, with a sly grin, Laris replies that that's just what the Romulans want the Federation to believe. She holds the device and makes a sweep across the room, the device emitting a thin blue ray. When the ray hits the couch, a visual reconstruction of the past shows Dahj and Caler talking, moments before Caler's death. The device is so clever that even the music that was playing in the apartment at the time can be heard.

So many questions are lit in our minds when we watch this scene in Dahj's apartment. How exactly does that Romulan forensics device interrogate recent history? How could molecular reconstruction possibly visually reenact the conversation between Dahj and Caler from the past? It seems incredible to us. And yet, when you think about it, so much of our science does something similar, though far less immediately.

LOOK-BACK DEVICES IN SCIENCE

Many modern sciences are historical in nature. Each uses its own look-back devices to interrogate the past. And each produces its own set of equally amazing, and sometimes questionable, results. The reading of starlight in astronomy and cosmology. The fossil record in paleontology. The periodic table in chemistry. The radiometric dating in geology and physics.

The curious thing is that most folks don't think of science as being historical. Maybe they swallow those ignorant words of American industrialist Henry Ford, about history being bunk. They subscribe to the rather ahistorical philosophy that current knowledge is the best available wisdom on science, that it has somehow replaced and supplanted all preceding scientific knowledge. But according to this belief, current knowledge too will become obsolete, displaced by future facts. All useful previous knowledge is subsumed by that of the present; the mistakes of the ignorant, consigned to the dustbin of history.

It's rather ironic, this ahistorical tendency. Especially when you think of notable revolutionaries like Darwin and Galileo. (Some readers may recall that the Galileo shuttlecraft was the first created for *Star Trek*, although a shuttle in *The Original Series* episode "The Menagerie, Part I" was the first actually televised shuttle, as that episode was aired before "The Galileo Seven.") The astronomy of Galileo and the evolutionary biology of Darwin are historical sciences. The vital significance of the theory of evolution was that it introduced a historical dynamic into fields of science. It broke with the orthodox branch of the Greek tradition, the eternal "truths" of Plato and Aristotle. And it returned to an earlier and heretical branch of the old Greek philosophers, the Atomists, with their focus on rational development and change.

EVOLUTION IS LIFEBLOOD

The rise of evolution injected the lifeblood of history into science. "He who . . . does not admit how vast have been the past periods of time may at once close this volume," Darwin wrote in his *Origin of Species*. For species to have evolved, the genuine extent of the Earth's past had to be much longer than the six thousand years or so suggested by biblical scholars. While biology and geology implied the Earth was ancient, however, they didn't prove it.

The spirit of evolution was gifted to the physicists. They too approached questions of history such as the age of the Earth, and the age of the Sun. In short, how old is time? And how have all things come to be as they are? First, the physicists used thermodynamics, that branch of physics concerned with the dynamics of heat energy. Then, late in the 1800s, the nuclear age dawned. Radiometric dating, the technique for dating materials using naturally occurring radioactive isotopes, provided age-dating to fields as diverse as geology, astrophysics, and cosmology.

By the 1920s, it was becoming clear to astronomers and geologists that the Earth was billions of years old. Rocks that were brought back by Apollo astronauts from the Moon, that natural satellite Galileo had spied through his new telescope centuries before, were dated at around 4.6 billion years old. And a consideration of nuclear matter in motion led astrophysicists to the conclusion that the Sun, which Galileo had observed to have spots and impurities, is a normal star, about halfway through its ten-billion-year evolution.

Most rational folks today "admit how vast have been the past periods of time," as Darwin so nicely put it. So let's look at a couple of choice disciplines of historical science and their associated look-back devices. How does each device work? And how exactly does it interrogate history?

Topic: Cosmology
Look-Back Device: The spectroscope

Stargazing is time travel. And gazing at stars helps us get a grasp of the sheer scale of the Universe. Since light, the fastest thing known to science, takes time to make any journey from one star system to another, it makes sense that light also takes time to cover the vast distances involved in a journey through space. For example, the nearest star to Earth, Proxima Centauri, is about 25 trillion miles away. Light takes more than four years, from that nearby star, to hit the naked human eye. Astronomers say that Proxima Centauri is four "light-years" away, since that's how long it takes light to make the journey. We are looking at a four-year-old image of the star.

At any moment, we see the sky as it was in the past. The farther we look out into space, the farther we look back in time, so it follows that we can use light to scale space and get a measure of the Universe. A light journey from the Moon to the Earth takes light around one second. We see the Moon as it looked a second ago. The Moon is a "light-second" away, we say. Similarly, a light journey from the edge of our solar system to the Earth takes around one light-year. We are between 24 and 28 thousand light-years away from the center of our Galaxy, and one of our nearest galactic neighbors, the Andromeda Galaxy, is about two and a half million light-years away. We see Andromeda as it appeared two and a half million years ago because it takes light that long to make its journey to Earth.

There's far more to be gleaned from starlight—consider the fascinating history of the spectroscope. It begins with English physicist Isaac Newton. By the time he was in his twenties, Newton became the first human to decipher the mystery of the rainbow. Using a prism, Newton confirmed that white light is a mixture of all the colors of the rainbow. And using a second prism, he showed that those colors can be reconstituted back into "white" light. (If you've ever wondered at the gatefold sleeve of the vinyl recording of Pink Floyd's *Dark Side of the Moon*, wonder no more. It's an illustration based on Newton's groundbreaking experiment that Sunlight is a blend of different colors and can be recombined again to make Sunlight.)

And yet, Newton missed something vital. Something maybe even more amazing that was hidden in the Sunlight. A kind of code. A cypher that proved to be a key to the cosmos. That code was later uncovered by Joseph Fraunhofer, a twenty-seven-year-old Bavarian and the world's leading designer of lenses, telescopes, and other optical devices. The code Fraunhofer uncovered led to the wedding of physics and astronomy, and the birth of the field of astrophysics.

Written in Sunlight is a kind of secret code in the form of a "pattern" of vertical black lines. Fraunhofer wrote, "I saw with the telescope an almost countless number of strong and weak vertical lines, which are darker than the rest of the color-image. Some appeared to be perfectly black." Fraunhofer mapped scores of the vertical lines in the Sun's spectrum. And he found the selfsame spectral patterns in the light of the Moon and the planets, which made sense, as these heavenly bodies shine simply by reflecting the Sun's light. But when Fraunhofer turned his telescope to the stars, the spectral lines looked very different. The importance of this difference remained a mystery for some time.

The spectral lines perplexed Fraunhofer. What was their origin? It must have been truly alien, if not diabolical, to see such dark, vertical lines set against the colorful glow of heavenly light. What was their message? It took a century of thinking, questioning, and research to decipher them, but Fraunhofer's lines are the atomic signatures of the elements writ large across deep space.

Stars like the Sun radiate light of all colors. If you look at that light through a prism, you see their spectrum, as Newton did. But when you magnify the spectrum with a spectroscope, as Fraunhofer did, you uncover evidence of an atomic dance. Evidence that an electron in an atom in a star has absorbed enough light to leap to a larger orbit. And that leaves a dark gap or black vertical line in the spectrum. Some of the dark lines in the spectrum will be the shadows cast by hydrogen atoms in the atmosphere of the Sun, for example. A different pattern of shadows will be cast by sodium atoms, as the electrons of other elements dance to different tunes. For example, every atom of iron has twenty-six electrons, and so the potential of their dance is so much more.

When you "read" starlight with a spectroscope, you're able to see the dark lines from all the elements in a star's atmosphere. Just think of the power of that. Given the spectrum of the Sun or a distant star, astronomers can tell what that star is made of. In the late 1800s, French philosopher Auguste Compte said that some things in the Universe are essentially unknowable. One of his main examples was the internal composition of the stars. Bad choice. The work started by Fraunhofer proved Compte wrong. Fraunhofer's lines not only show that the atomic signatures of the elements are writ large across the Universe, but the dark spectral lines can also be used to conjure data on the temperature, composition, and movement of stars, and even provide evidence for the birth of the cosmos itself.

That's where the famous "red shift" comes in. The spectroscope and the dark Fraunhofer lines are a look-back device that provides the key evidence of the Big Bang itself. When we look at distant Galaxies, we're looking way back into the past. Galaxy light takes billions of years, longer than the age of the Earth, to reach our spectroscopes. The speed with which Galaxies race apart can be estimated from their effect of shifting the Fraunhofer lines toward the red end, hence the famous red shift. That's because a moving source of waves causes a "Doppler shift" of the Fraunhofer lines. (Something similar happens to sound waves when a car engine's pitch is distinctly higher when approaching than when receding.) Speeding light waves Doppler shift the Fraunhofer lines to the blue end of the spectrum, if they're moving very fast toward us, and to the red end of the spectrum if they're moving very fast away from us. That's true of distant Galaxies. And Fraunhofer's dark lines are not only evidence that the Galaxies are fast receding but they also give an estimate of when the Universe began expanding.

Topic: Chemistry

Look-Back Device: The periodic table

Okay, so some of you may have loathed the periodic table back in your school days. You know the tale. Back in Ancient Greece, even the best philosophers thought that the cosmos had but five elements. Four of them, Earth, air, fire, and water, were to be found on the Earth only. And the fifth element, the quintessence, was only to be found in the heavens. Today, we know of 118 elements. Some have been known since before written history, such as gold (Au), silver (Ag), copper (Cu), lead (Pb), tin (Sn), and mercury (Hg), while elements 99 to 118 have only been relatively recently synthesized in laboratories or nuclear reactors.

The key to understanding the elements is the periodic table, a pattern embedded in nature, which miraculously came to Russian chemist Dmitri Mendeleev in a dream: "I saw in a dream, a table, where all the elements fell into place as required. Awakening, I immediately wrote it down on a piece of paper."

Did you know that the periodic table was also a look-back device? Its status as such is related to the history of astronomy detailed above. Once astronomers had evidence of red shifts and the Big Bang, they realized that the Universe is evolutionary. And this evolutionary nature also applied to the chemical constituents of the cosmos. As famous cosmologist George Gamov put it, "We conclude that the relative abundances of atomic species represent the most ancient archaeological document pertaining to the history of the Universe."

In other words, the periodic table is evidence of the evolution of matter, and atoms can testify to the history of the cosmos. And yet, early versions of Big Bang cosmology supposed that all the elements of the Universe were fused in one fell swoop. As Gamov put it, "These abundances must have been established during the earliest stages of expansion, when the temperature of the primordial matter was still sufficiently high to permit nuclear transformations to run through the entire range of chemical elements."

It was a neat idea, but very wrong. Only hydrogen, helium, and a dash of lithium could have formed in the Big Bang. All elements heavier than lithium were made much later, by being fused in evolving and exploding stars. How do we know this? Because at the same time some scholars were working on the Big Bang theory, others were trying to ditch the Big Bang altogether. Its association with thermonuclear devices made it seem hasty, and its implied mysterious origins tainted it with creationism.

And so, a rival camp of cosmologists developed an alternate theory: the Steady State. The Steady State held that the Universe had always existed. And always will. Matter is created out of the vacuum of space itself. Steady State theorists, working against the Big Bang and its flaws, were obliged to wonder where in the cosmos the chemical elements might have been cooked up, if not in the first few minutes of the Universe. Their answer: in the furnaces of the very stars themselves. They found a series of nuclear chain reactions at work in the stars. First, they discovered how fusion had made elements heavier than carbon. Then, they detailed eight fusion reactions through which stars convert light elements into heavy ones, to be recycled back into space through stellar winds and supernovae.

And so, it's the inside of stars where the alchemist's dream comes true. Even every gram of gold began billions of years ago, forged out of the inside of an exploding star in a supernova. The gold particles lost into space from the explosion mixed with rocks and dust to form part of the early Earth. They've been lying in wait ever since. All of which means that the periodic table is a look-back device. It's a phylogeny of matter, evidence of the history of the evolution of chemical species, and an incredible documentation of the history of the cosmos.

PART III
MACHINE

WHAT *STAR TREK* MACHINES CAME TRUE?

"Almost all of this comes out of my feeling that the human future is bright. We're just beginning. We have wonders ahead of us. I don't see how it can be any other way, with the way the future is going. We now have got a telescope up there. We're photographing the Universe. We're inventing the next life form, which is the computer. We're in the midst of it. And it will happen."

—Susan Sackett, personal conversations
with Gene Roddenberry (1990)

FUTURE DREAMING

Some Trekkers are not happy. The *Star Trek* future we were promised hasn't really materialized, they say. According to these Trekkers, we're now meant to be living in a warp-speed world. A world also of color-ranked jumpsuits, phasers, and com-badges. Maybe the aliens are also meant to have landed. Some Trekkers go so far as to blame twenty-first-century science and tech. Why can't they keep pace with the fiction? It's an interesting question.

Given that the Star Trek franchise is based on a mission to explore strange new worlds, to seek out new life and new civilizations, let's go back to the original idea of those "new worlds" in the first place. The following example from the history of science may help answer the Trekkers' question as to why "fact" can't keep up with ideas.

The existence of exoplanets in orbit around other stars was first suggested by Greek Atomists. In the fifth century BC, the philosopher Democritus, after whom democracy is named, believed in a kind of cosmic democracy. Though Democritus didn't call them "exoplanets," he nonetheless correctly said that Suns and planets were everywhere in the cosmos.

Naturally, the mere thought of the idea of exoplanets wasn't enough to make them materialize. That didn't happen until 1995. And the tech that ultimately kick-started the discovery of exoplanets was the invention of the telescope, which was way back in 1600, more than two thousand years after Democritus. In a very real way, the telescope, also known as the far-seer, brought the heavens down to Earth. The technology that made up the far-seer had long been familiar. But the history of the optics of the telescope includes a complex social relationship between theory and technique. Ancient civilizations had noticed the curious lens effect produced by, say, transparent crystals, glass spheres full of water, or jewels. Indeed, the magnifying effects were something of a fascination.

But it's as though those ancient observations came too soon. In the theories and texts of many ancient Greek and medieval Arab scholars, it's clear that a theory of optics existed. The tech to magnify had also been available. And clearly the demand for the tech of magnification in commerce and warfare would have been profound. But theory was never married to practice, never tallied up to the optical devices that may have been available.

This example of tech lagging behind ideas can be explained in two related ways. First, a social explanation. It's very unusual in history for scholars and craftsmen to work together, given the difference in their social class. This was the case until the European Renaissance, one of the distinguishing features of which was the intensity and distribution of the collaboration between the work of the mind and the work of the hand.

Second, the old adage "a little knowledge is a dangerous thing" may also apply. Ancient or medieval observations made through transparent crystals, glass spheres, or jewels don't just show things bigger or better. They can also deceive. To the more superstitious mind, a mind not married to rational theory, such optical illusions can make one wary. Sight is the most paradoxical of senses. It is at once the most reliable, and yet the most deluding. Without theory, perhaps you can't really trust what you see.

BE THANKFUL

What does the almost two-and-a-half-thousand-year delay between the idea and actual discovery of exoplanets have to do with *Star Trek* tech? Again, the telescope is key. The example of the exoplanets shows that cultural ideas run ahead of technical invention and discovery. Thought and theory is one thing. But practical tech is needed to realize that theory. And the tech of a particular time period is linked to the material progress of society. In short, we can't be sure of exoplanets until we have the technical means to know that they're out there. And that takes time.

Having said all that, surely there's some tech from *Star Trek* that has made its way into the markets of commodity fetishism. Surely this iconic franchise, which has been a cult phenomenon for decades, has generated the odd invention or gadget idea that has influenced actual modern tech. Yes, of course, so let's be thankful for small mercies that we won't have to wait two-and-a-half-thousand years before we see *Star Trek*'s influence on late-capitalism!

THE TRICORDER

In the Star Trek Universe, the tricorder is a multifunctional handheld device used for environment scanning and data recording and analysis. (Indeed, the very word "tricorder" is short for "TRI-function reCORDER," which refers to the device's three main functions: sensing, recording, and computing.) Three main types of tricorders appeared in *Star Trek*. First, the standard tricorder. This is a general-purpose handheld, used mostly to scout unfamiliar settings, render in-depth examinations of living things, and record technical data. Second, the medical tricorder. This is used by medical staff to help in the diagnosis of diseases and to gather bodily data about a patient. Third, the engineering tricorder is calibrated for starship engineering functions. There are also many other lesser-seen types of bespoke tricorders.

The tricorder as seen in *The Original Series* was a black, rectangular affair. It had a top-mounted rotating hood along with two opening compartments and a shoulder strap. The top was able to pivot open, revealing a small screen with control buttons. McCoy used an adaptation of this model, which had a detachable "sensor probe" stored in the bottom compartment when not used. The design team used a probe that was fashioned from a saltshaker to save costs. The tricorder prop for *The Original Series* was designed and built by Wah Ming Chang. Chang created a number of futuristic props as part of his contract, and some of his designs were influential on later, real-world electronics devices. For example, NASA uses a handheld device known as LOCAD. Somewhat like the standard tricorder, LOCAD detects unwanted micro-critters such as E. coli, fungi, and salmonella on board the International Space Station.

What's more, two handheld clinical devices, similar to the medical tricorder, may soon help doctors examine blood flow and check for bacterial infection, diabetes, and even cancer. Medical researchers and scientists at Loughborough University in England have developed photoplethysmography (a simple optical technique used to detect volumetric changes in blood in peripheral circulation) in a handheld device, which is able to monitor the functions of the heart. Meanwhile, researchers at Harvard Medical School have made a small device that uses similar tech found in MRI machines that noninvasively inspect the body. The device uses nuclear magnetic resonance (NMR) imaging. NMR is a physical phenomenon in which nuclei in the strong constant magnetic field of the machine are perturbed by a weaker oscillating magnetic field and respond by producing an electromagnetic signal pattern. This Harvard device will be sensitive enough to measure samples of as many as ten potentially infectious bacteria. This type of sensitivity, eight hundred times more sensitive than the traditional kit used in medical labs, could lead to a revolution in the way doctors diagnose disease.

COMMUNICATORS

Ever watched an old movie or TV program and imagined how people could have coped better if they had cell phones? For example, in *Goldfinger*, Bond would have had advance warning of Odd Job's deadly hat, as Q would have seen a similar hat for sale while browsing through eBay. In *Jaws*, Chief Brody could have easily sorted out his problem with the humongous but dumb great white by simply organizing a flash mob of resourceful, experienced shark hunters on Twitter. And in *Psycho*, before checking into the Bates Motel in a deserted California backwater, Janet

Leigh could have simply looked up the joint on Trip Advisor and found the following entry: "Stay well away. Creepy owner, constantly talking to a seemingly absent mother. Filthy shower, and no Wi-Fi."

Well, there were never communication problems in the Star Trek Universe. From the very early days of *The Original Series*, Captain Kirk was sorted. Sure, whenever he left the safety of his ship, it could well be the last time he saw it. Danger is rarely far away when you're seeking out new life and new civilizations and so on. But our good Captain could count on it that, whether he needed Scotty to beam him up, Spock to feed him some vital intel, or Bones to provide some futuristic potion, Kirk could simply whip out his communicator and the deed was done.

Fast-forward to today and almost everyone carries a communicator of sorts. According to BankMyCell.com, "In 2021, the number of smartphone users in the world . . . is 3.8 billion, which translates to 48.53% of the world's population owning a smartphone. In total, the number of people that own a smart and feature phone is 4.88 billion, making up 63.60% of the world's population." Of course, we can nitpick that the communicators in *Star Trek* were not really smartphones. They were more like the push-to-talk, person-to-person devices first made popular by Nextel in the nineties. The *Star Trek* communicator had a flip antenna that opened and activated the device. We might argue that the original flip phones were distant cousins. (But let's be clear, the creators of the *Star Trek* communicator imagined a device for voice communication, which most certainly served as a muse for real-world technology products.)

Later versions of the *Star Trek* communicator evolved into the Starfleet com-badges, a logo with a comms link located on a crew-member's chest. On the mere tap of a finger, communication between crew members became easier still. And this more refined communicator also acted as a muse for real-world tech. To take just one example, Vocera Communications made a com-badge that can link people on a network inside a designated area, such as an office or a building, by using the badge's software over a wireless LAN. The B2000 com-badge allows clear two-way communication, weighs less than two ounces, and can be worn on a lapel. The B2000 is even designed to limit the growth of bacteria, making it suitable for medics. In December 2016, Fametek also released a com-badge known as the Star Trek: The Next Generation CommBadge. It uses Bluetooth 4.2 tech, which enables the badge to pair with Bluetooth enabled cell phones or tablets, and can receive and make calls or use voice commands with Google, Siri, or Cortana. Even better for some Trekkers, this Bluetooth ComBadge has a cosplay mode that activates when pressed and makes the same chirp sound effect that's seen on the show.

The *Star Trek* communicator goes beyond mere cell phone capabilities, even if they went on to inspire cell phone prototypes. For instance, the communicator enables crew members to contact orbiting starships without the need for an artificial satellite to relay the signal. The communicators appear to make use of subspace transmissions that don't conform to usual laws of physics—they can bypass electromagnetic interference. The devices also appear to allow almost instantaneous communication at distances that would otherwise require more time to cross. In the real world, no equivalent to subspace communication has been theorized, let alone developed. But many other aspects of Starfleet comms tech have been developed. In particular, the locator/transponder functionality exists in forms such as GPS and radio direction finder devices. To some extent, it's also true that cloud-based digital assistants work in a way that is similar to the artificial intelligence of a Starfleet ship's computer.

TELEPRESENCE

Back in 1966, the idea of interacting with another human who was situated elsewhere in spacetime seemed science fiction indeed. Telepresence is the set of technologies that enables a person to feel as if they were present, via tele-robotics, at a place other than their actual location. The idea of telepresence sprang up in sci-fi a generation earlier than *The Original Series*. According to legendary American cognitive scientist Marvin Minsky, the origin of the telepresence concept was down to US science fiction author Robert A. Heinlein. Minsky said that his own vision of a remote-controlled AI economy came from Heinlein's prophetic 1942 novel, *Waldo*, in which Heinlein proposed a primitive telepresence master-slave manipulator system.

In *Star Trek*, the idea of telepresence was extended to a process that made it possible for human, alien, or android to safely control a vessel from the safety of a remote location. The idea of telepathy is related to telepresence in the sense that it was often a required ability to successfully use a telepresence unit. In reality, it's a requirement of telepresence that the sense of the user be stimulated in such a way as to give the feeling of being in that other location. What's more, users can be given the ability to alter aspects of this remote location. In some cases, the user's voice, position, actions, or movements can be sensed, relayed, and copied in the remote location so as to bring about some desired effect. Thus, data may be traveling in both directions between the user and the remote location.

As early as 2008, the world's largest telecommunications company AT&T teamed up with technology conglomerate Cisco to deliver the first in-depth telepresence experience. Cisco's TelePresence combined audio, video, and ambient lighting to mirror the surroundings and mimic sounds so that users on each side of the conference felt as though the images on the screen were in the same room as them. For example, people in boardroom Q, let's say, would see people on the screen in boardroom K as though they were sitting across the table from them. Aspects of ambient lighting and room layouts were configured to mirror one another. While Cisco's TelePresence kit was maybe more advanced than implied by *The Original Series*, there's little doubt once more that the franchise was the muse of the tech.

UNIVERSAL TRANSLATOR

Surely one of sci-fi's most deliciously comic inventions is Douglas Adams's Babel fish. The Babel fish was a fictional species of fish invented by Adams in his 1978 work *The Hitchhiker's Guide to the Galaxy*. When placed in your ear, the Babel fish could translate any tongue simply by eating the language spoken to you and spitting out a translation in your own language into your ear.

Something similar, though much less exotic, existed in *The Original Series*. No matter what planet you visited, no matter how alien the culture, you could understand all that the indigenous people were saying. What a travel essential that would be. And that's exactly what the planet-hoppers Kirk and crew had onboard the Enterprise—a universal translator. The crew members in *Star Trek* were kitted out with a small gadget that, when spoken into, would translate the words into English. And yes, this franchise muse has also become tech reality. Today, there are translation devices that let you speak phrases at them and will answer back in the language you want. The trouble is, of course, unlike the Babel fish and the *Star Trek* universal translator, these translation devices only work for a set of programmed languages. Having said that, the technology now exists for fact to match the fiction, as long as that fiction isn't as fantastic as the Babel fish. Voice recognition has advanced hugely since the days of *The Original Series*. But computers still have some way to go before they are able to actually learn languages. Sure, computers can theoretically gather data faster than the human brain, but programs are dependent on software. Some human has to spend the time, and money, to compile a universal translator, which is probably why current systems focus on more popular languages. For now.

PHASERS

Like "live long and prosper" and "beam me up, Scotty," "set phasers to stun" is one of those *Star Trek* quotes that might just help you go boldly through your day. The Enterprise crew often depended on the stun setting of the fictional weapon known as the phaser. Indeed, forearmed with a phaser, Kirk and co had the pacifist option to stun their opponents, rendering them incapacitated, rather than kill them outright. From *The Original Series* on, phasers have been a common and convertible weapon of choice, seen or referenced in almost all franchise film and television titles. The phaser comes in a broad range of sizes, from small handheld guns and rifles to starship ordnance. Though they appear to fire a beam, phasers in fact discharge a steady stream of pulsed energy projectiles at their target.

Now, stun guns have been around for some time. For example, in 1935 Cuban inventor Ciril Diaz designed an electroshock glove. But, three years after *The Original Series* first aired, Jack Cover, a researcher with NASA, began developing the taser. By 1974, Cover had completed the device, which he named after his childhood hero Tom Swift, a character from a series of American juvenile sci-fi novels, which showcased science, invention, and technology. Thus, taser stands for "Thomas A. Swift's electric rifle." Cover's model, known as the Taser Public Defender, used gunpowder as its propellant, which in 1976 led the Bureau of Alcohol, Tobacco, Firearms, and Explosives to classify it as a firearm. But in 1983, Cover's patent was developed by Nova Technologies into the Nova XR-5000—a non-projectile handheld-style stun gun. It was the XR-5000 design that was widely copied as the progenitor of the compact handheld stun guns used today. The taser may fail to kill like the phaser, but it packs a sufficient electrical punch to render its victim stunned, if not completely incapacitated.

Of course, a taser is unlike a phaser in many ways. For example, stun guns, like the taser, work by coming into actual physical contact with the target. It's how they stun. They fire a pair of electrodes connected by wires that, unfortunately for the target, attach to the body. Once contact is made, the gun sends out a jolt of electricity to the target, thus the stun. There's also an inferior type of stun gun. They have stationary electrical contact probes and pack less punch in the sense that, while they have a similar effect on the target, the gunman has to be within arm's length to be able to zap their target!

Will phasers soon hit grocery store shelves? Will gun fetishists soon be able to pick up a weapon which discharges a stream of pulsed energy projectiles into a target? Well, something more along phaser lines could be in development. A company called Applied Energetic has developed technologies, known as Laser Induced Plasma Energy and Laser Guided Energy, that reportedly transmit high-voltage bursts of energy to a target. In short, the energy pulses would stun the target and limit collateral damage, which means a true phaser may soon be a reality.

TRANSPARENT ALUMINUM ARMOR

Earlier in this book we spoke in some detail about the plot of the *Star Trek IV: The Voyage Home* movie in which Kirk, Spock, and co travel back in time to rescue some humpback whales. When our heroes return to present-day Earth, they junk a Klingon Bird of Prey in San Francisco Bay after almost hitting the Golden Gate Bridge while flying blind in a storm.

This scene is a memorable one. But perhaps a little less memorable is the fact that Scotty showcases transparent aluminum for the first time. To be clear, Scotty trades in sheets of plexiglass for the formula matrix for transparent aluminum, which represents huge engineering progress. Scotty's aim is to build a tank to transport the two humpback whales. And his trade claim was that his "buyers" would be able to replace six-inch-thick plexiglass with one-inch-thick transparent aluminum.

While Scotty's trade-off sounds fanciful, there *is* such a thing as transparent aluminum armor, or aluminum oxynitride (ALON), as it's better known. ALON is actually a ceramic material. It begins life as a powder which, when subjected to heat and pressure, transforms into its crystalline form, similar to glass. In its crystalline form, ALON is tough enough to brave bullets. Polishing the molded material strengthens it even more. Research engineers and scientists have tested the ALON material in the hope that it can substitute windows and canopies in high-speed aircraft. After all, transparent aluminum armor is stronger and lighter than bulletproof glass—it's almost a done deal.

STAR TREK:
WHEN WILL WE BOLDLY GO?

"Why a journey into space? Because science is now learning that the infinite reaches of our Universe probably teem with as much life and adventure as Earth's own oceans and continents. Our Galaxy alone is so incredibly vast that the most conservative mathematical odds still add up to millions of planets almost identical to our own—capable of life, even intelligence and strange new civilizations. Alien beings that will range from the fiercely primitive to the incredibly exotic intelligence which will far surpass mankind."

—Gene Roddenberry, *The Hollywood Reporter*
(September 8, 1966)

"The exploration of space will go ahead, whether we join in it or not, and it is one of the great adventures of all time, and no nation which expects to be the leader of other nations can expect to stay behind in the race for space."

—John F. Kennedy, speech at Rice University
(September 12, 1962)

THAT FINAL FRONTIER

One evening in September of 1966, Americans sat down to watch the first episode of a brand-new science fiction television series. That same month, inventor Ralph H. Baer began work on the basic principles for creating a video game to be played on a TV set; the US Defense Department demanded the largest draft call of the Vietnam War, calling for almost fifty thousand men to be inducted into military service; and NASA was finalizing its plans to launch its ninth two-man trek into "the final frontier," this time with the Gemini 11 mission.

Gemini, with astronauts Charles Conrad and Richard Gordon, was a little over twelve hours from take-off when *Star Trek* premiered on NBC. And, like their fictional TV counterparts, Gordon and Conrad were also set to "boldly go," as Gemini 11 was to reach an altitude of 850 miles, almost twice as far into space as any man had gone before. (Technical hitches meant that Gemini's launch was delayed a few days, but their three-day mission remained—to seek out a rendezvous target and explore the vacuum of space.)

That early *Star Trek* audience was treated to a future based on the actual spaceflight endeavors of the NASA program. Subsequent releases of the franchise, TV programs and movies alike, would more obviously draw upon that link between science fact and fiction. The fictional Universe of the Starship Enterprise sometimes referred to actual NASA spacecraft and milestones.

Now that we are two decades into the twenty-first century, when will we finally boldly go? Heavens above, in the *Star Trek* timeline, Zefram Cochrane, inventor of the warp drive, is born in the year 2032; the first human deep-space colony, Terra Nova, is set up by 2069; and by 2103 Earth colonizes Mars! We need to get a move on. But, like the launch of Gemini 11, the human space program has been delayed.

In October of 2020, however, NASA revealed conclusive evidence of water on the Moon. Unlike previous evidence detecting water in permanently shadowed parts of the Moon's many craters, new evidence detected water molecules in Sunlit regions of the lunar surface. This finding prompted NASA to confirm it will send the first woman and next man to the Moon's surface in 2024.

TWENTY-FIRST-CENTURY SPACE RACE

Will this lead to a twenty-first-century space race? Could we soon see ten-minute space vacations, orbiting space hotels that only rich people can afford, or even humans on Mars? Beating the *Star Trek* prediction of Martian colonization by 2103 will take some organization! Unlike the past, when superpowers jostled for dominance in orbit with no obvious victor, private corporations are now vying to make space travel possible and affordable for the wealthy. For example, SpaceX recently completed the goal of launching humans to the International Space Station (ISS). But other missions will soon be on the cosmic horizon.

The USSR also made great strides into space. After all, the Space Age began in 1957 with the launch of *Sputnik*, Earth's first artificial satellite. The first living creature, the dog Laika, was also launched into space in 1957, followed by the first man in space, Yuri Gagarin in 1961, and the first woman, Valentina Tereshkova in 1963. And in 1965, Alexei Leonov became the first human to walk in space. There may be some truth to the USSR newspaper *Pravda*'s complaint that there were no Russians among the culturally diverse *Star Trek* cast. It was seen as a slight to a country that had produced so many firsts in spaceflight. Gene Roddenberry said in response, "The Chekov thing was a major error on our part, and I'm still embarrassed by the fact that we didn't include a Russian right from the beginning."

NASA's link with private corporations continues. For example, NASA is counting on Boeing and SpaceX to build spacecraft that can transport humans into orbit. Once built, both corporations keep ownership and control of the craft so that NASA can send people into space for a fraction of its usual cost. And other corporations too, like Virgin Galactic and Blue Origin, are specialists in suborbital space tourism. For example, a test launch movie, filmed from the interior of Blue Origin's New Shepard craft, boasts breathtaking views of the blue marble Earth below. It was a relatively calm journey for its first passenger, a bench test dummy dubbed "Mannequin Skywalker." Sadly, someone completely failed to exploit the more obvious links between space exploration and *Star Trek*, given *Star Wars* takes place in another Galaxy entirely. Another example of suborbital space tourism is Virgin Galactic's suborbital spaceplane. It will offer paying customers around six minutes of microgravity and weightlessness during its very brief journey through our planet's atmosphere. With these types of spacecrafts in development, countless dreams of spaceflight will be distilled down to only the very select few passengers able to pay the princely sums mulcted for the experience.

STEPPING-STONE MOON

For many years, humans have seen the Moon as a stepping-stone in the exploration of more distant worlds. After a long absence from the lunar neighborhood, NASA is now planning to place a space station in lunar orbit sometime in the next decade. But, before that, the agency also aims to put the first woman, and the next man, on the Moon's surface by 2024 through the Artemis Program, a sister to the Apollo missions of the 1960s and 1970s.

China's government is also planning its own mission to the Moon, Mars, and beyond. For one thing, the Chinese plan to build a lunar space station. They also plan to set up a scientific outpost at the south lunar pole and send astronauts to land on the Moon. The so-called International Lunar Research Station would be situated at the south pole of the Moon. European space agencies, along with the Russians, are considering cooperating with the development of ILRS. The aim is to have a robotic presence at the pole, along with short-term human missions there in the 2030s. Longer-term human habitation at the ILRS could begin in 2036.

Longer-term lunar habitation builds both the experience and expertise humans need for the protracted space missions necessary to visit the other planets in our solar system. The Moon may also be used in the future as a forward base of operations. From the lunar surface, we may learn how to replenish essential supplies, like rocket fuel and oxygen, by manufacturing them from local material.

Skills such as these are vital for the expansion of human civilization into deep space. A Moon presence makes us less tied to Earth-based resources. Even though we have visited the Moon before, the lunar surface still bears its own scientific mysteries. Missions will explore the extent of water ice near the south lunar pole, which is why that particular location is one of the top target destinations for space exploration.

ORION

The focus of the Artemis program is a new, state-of-the-art spacecraft called Orion. Like the spacecraft of Russia's Soyuz spacecraft, as well as the Mercury, Gemini, and Apollo programs, Orion is a pretty standard space capsule. But it's larger than those that went before and can accommodate a four-person crew. Some science fiction fans may be somewhat disappointed by Orion's retro design. But the capsule concept is thought to be more reliable and far safer than the more revolutionary space shuttle designs, such as the Russian Buran and NASA's own shuttle, which couldn't travel beyond Earth's orbit and was prone to catastrophic failures.

Design engineers know that capsules can afford launch-abort capabilities. These can protect astronauts in the event of rocket malfunction. Furthermore, the mass and design of space capsules means they can travel beyond Earth's atmosphere.

NATIONAL OR INTERNATIONAL

NASA is taking a decidedly national approach to space exploration. It hopes that its partnerships with corporations will enable it to reach the lunar surface by changing spaceflight economics through increased competition that drives down costs. The hope is that space travel will become cheap enough so that private citizens don't have to be hugely wealthy.

But *Star Trek* rarely portrayed space exploration as a way to promote national pride. The franchise saw the future of space as international. International partnerships broaden our knowledge of the cosmos. They help in the search for life on other planets. They aid in our vital observations of our home world. And they help in the swifter movement of humans into space over time.

Developments in science and tech are not the sole preserve of one country alone. It's diverse and creative teams that solve problems. As American space scientist James Van Allen put it, "Outer space, once a region of spirited international competition, is also a region of international cooperation." Allen says he realized this as early as 1959, when he attended an international conference on cosmic radiation in Moscow: "At this conference, there were many differing views and differing methods of attack, but the problems were common ones to all of us and a unity of basic purpose was everywhere evident." Allen concluded that "Many of the papers presented there depended in an essential way upon others which had appeared originally in as many as three or four different languages. Surely science is one of the universal human activities." Marie Curie said much the same thing: "After all, science is essentially international, and it is only through lack of the historical sense that national qualities have been attributed to it." No one country or corporation can go it alone when it comes to the final frontier of space.

HOW MIGHT SPECIES 8472 DESTROY A BORG CUBE IN SECONDS?

"I am not only a pacifist but a militant pacifist. I am willing to fight for peace. Nothing will end war unless the people themselves refuse to go to war. He who joyfully marches to music in rank and file has already earned my contempt. He has been given a large brain by mistake, since for him the spinal cord would suffice."
—Albert Einstein, *The World as I See It* (1931)

"Sgt. Brad 'Iceman' Colbert: All religious stuff aside, the fact is, people who can't kill will always be subject to those who can."
—David Simon, *Generation Kill* (2008)

"I know not with what weapons World War III will be fought, but World War IV will be fought with sticks and stones."
—Albert Einstein, interview with Alfred Werner,
Liberal Judaism (1949)

DUCK EATS YEAST, QUACKS, EXPLODES; MAN LOSES EYE

Consider first an explosion associated with another species. On Earth, in January 1910, dozens of newspapers across the United States reported one of the strangest ever accidents in local history. The event appears to have happened in Des Moines, Iowa, when a duck by the name of

Rhadamanthus won a prize at an Iowa poultry show. By all accounts, Rhadamanthus then promptly exploded into smithereens, one of which struck a man named Silas Perkins in the eye, thus ruining his sight on that side.

What was the cause of such an unlikely explosion? Apparently, his duck Rhadamanthus (named after the mythical king of Crete) had partaken of some yeast that had been placed in a pan on the back porch. During his Sunday morning stroll, the soon-to-be-ex-duck had been tempted by the yeast and pecked at this impromptu feast. On returning from church, Rhadamanthus's owner, said Silas Perkins, found his prize duck in a somewhat sluggish mood. Telltale marks about the pan of yeast raised Silas's suspicions about the cause of the duck's condition. Perkins was about to pick up the duck when the latter quacked, then exploded with a loud report into smithereens. Silas ran into the house, holding both hands over one eye. A doctor was promptly called, but he found that the eyeball had been so profoundly penetrated by a fragment of flying duck that there was no hope of saving the embattled optic.

THE BORG

So, what's the link between the exploding duck and species 8472 destroying a Borg cube in seconds? In short, it's the improbability. The main "improbability" we shall look at here is the apparent vulnerability of the Borg cube, and how it might have been so easily exploded by species 8472. You can work out for yourself whether it is the Borg or species 8472 which is Rhadamanthus, and which is Silas Perkins. But, first, consider some other improbabilities. As we all know, the Borg first rose to prominence in the 1989 *Star Trek: The Next Generation* episode "Q Who." The Borg quickly became the most popular villains on *Star Trek*: a cyborg species with a charming and unrelenting mission to indoctrinate all life forms into their collective. No doubt, Seven of Nine aside, the Borg's popularity centered around the mystique of their advanced tech, their merciless strategies, and their carefully strict and tactical approach to all matters.

With that characterization in mind, why did the Borg make so many improbable and fatal blunders when it came to Species 8472? Not only did the Borg invade a space they were not meant to occupy, but they also attacked an alien race they knew nothing about and were ill-prepared for the ensuing combat. Okay, it does make some sense for the Borg to come across Species 8472 by chance and then react incompetently. To some extent. But seeking out such danger in the first place without recon or sufficient intel? Most inimical. The story would have made more sense if a third species had broken into the fluidic space of Species 8472 when the Borg had been pursuing this other aggressive species.

SPECIES 8472

Anyhow, what do we know of Species 8472 that the Borg didn't? We know they originate from what is called fluidic space. As the only sentient species native to this space, Species 8472 then launch a genocidal retaliation against all the weak and impure species of the Galaxy, including a systematic extermination of the Borg, as well as causing collateral damage of innocent species. Over the next half a year, tens of billions of drones are lost along with several hundred planets and ships. That's an expensive mistake to make.

What kind of scientific and technical capacity do Species 8472 possess? Well, while they may only be mostly known by their rather vague Borg designation, Species 8472 certainly seem to be a highly evolved biological terror, somewhat reminiscent of the primal, predatory creatures sometimes known as the xenomorph in the Alien franchise, but with much better tech. (The word xenomorph, which means "alien form" from the Greek *xeno*, which means "other" or "strange," and *morph*, which suggests shape, was first used by the *Aliens* character Lieutenant Gorman with reference to generic extraterrestrial life, but the word was mistakenly assumed by some fans to mean that specific creature in Aliens.)

Species 8472 are large tripedal beings with devastating claws and acute telepathic senses. The trait of telepathy has often been used in the history of sci-fi to suggest a highly evolved alien intelligence. Indeed, Species 8472 do appear to be highly evolved: they don't even need a breathable atmosphere for respiration or locomotion, and they seem to have developed an advanced bio-tech culture, which allows individual members of the species to integrate with their vessels. And we know that they are certainly advanced enough to not only overpower the Borg but were on the verge of winning a war against all Borg before Janeway intervened.

BORG CUBE

Species 8472 certainly seem technically sophisticated enough to blow up a Borg cube in seconds. Besides, we know that they did! Given that their target was a Borg cube, let's take a longer look at one. According to the *Star Trek: Voyager* episode "Dark Frontier," Borg cubes measured over three kilometers cubed, with a resulting internal volume of approximately 28 cubic kilometers. *Star Trek: Voyager* episode "Endgame" informs us that the cubes are constructed from an extremely hard alloy of the fictional tritanium (which is reportedly over twenty-one times harder than diamond), and the ore of which can be found on the planet Argus X.

Now, Arthur C. Clarke and Stanley Kubrick's movie *2001: A Space Odyssey* is still regarded as one of the most influential and important pieces of science fiction ever made. *2001*'s main spaceship, Discovery One, was designed with one major thought in mind: to accurately represent how space travel is actually engineered; in particular, the design of Discovery One and other craft was based on engineering considerations instead of lame attempts to look aesthetically "futuristic." Many other sci-fi movies give spacecraft an aerodynamic shape, which is superfluous in outer space. Kubrick knew better.

The same goes for the aerodynamically curious Borg cube. Indeed, consider the real-life CubeSat. The CubeSat is a U-class spacecraft, a kind of miniaturized satellite used for space research. A CubeSat is made up of multiple cubic modules, each ten centimeters cubed with a mass of around 1.3 kilograms. They're commonly sent into orbit by deployers on the International Space Station, and as of January 1, 2021, more than 1,350 CubeSats have been launched.

In the *Star Trek: The Next Generation* episode "The Best of Both Worlds," it's estimated that a cube could stay functional, even if over three-quarters of it was rendered inoperable in battle, due to the decentralized and redundant nature of its key systems. According to the *Star Trek: The Next Generation* episode "Q Who," Borg cubes were also very decentralized in structure. Unlike the Enterprise and other spaceships, the Cubes had no specific bridge to speak of, nor living quarters or engineering sections. All such systems were distributed throughout the ship, which, along with the presence of a regenerative hull, made it somewhat resistant to damage and system shutdowns.

Another aspect of Borg cube construction will shortly become relevant in our consideration of blowing one up: the cubes are very roomy affairs. According to the *Star Trek: Voyager* episode "Dark Frontier," Borg cubes are more like traveling city-ships than starships. Despite their size, around half of their internal volume is empty space. Only a staggering 0.2 percent of its internal mass is actual walls and floors. Situated between such boundaries are alcoves, corridors, hangar bays, and other functional spaces.

BLOWING UP THE BORG

Let's consider that the preferred method which Species 8472 use on the cube is to smash it to smithereens like poor Rhadamanthus. What size of explosion would do that? (I am going to deliberately avoid silly super-weapons in this solution. You know the kind of thing; weapons in superhero sci-fi in particular have daft names like the Mobius Chair, the Helmet of Fate, or the Heart of the Universe. Instead, we shall base our preferred method of attack on actual projected science.)

So, let's smash the Borg cube to smithereens. The gravitational binding energy of a system like the Cube is the minimum energy that must be added to it in order for the system to cease being in a gravitationally bound state. In other words, the binding energy of a Borg cube is that energy which stops it from separating into smithereens. Now, calculating the binding energy of a sphere or cube of uniform density is tricky enough. But, as we have just discussed, the Borg cube is not of uniform density. As a compromise, because a Borg cube is half empty space, let's use a (very) rough estimate that the approximately 28 cubic kilometers behaves like a sphere of half that volume. A Borg cube has a mass of 90,000,000 metric tons, or 90,000,000,000 kilograms. So, when we plug our values into the necessary equation (the binding energy is proportional to the mass of the cube squared, divided by 5 times the Cube's "radius"), we get an answer of

216,237,600 joules of energy, or roughly 216 and a quarter million joules. That's a pretty impressive-looking bundle of energy.

Exactly how impressive a bundle of energy is it? Well, an ordinary car, speeding along a highway, has an energy density of about 360 joules for each kilogram. For instance, a fully loaded Toyota Corolla (my old car) would have a total energy of around 612,000 joules, the energy it would hit you with while coming to a complete stop in the process. Meanwhile, the cruising speed for a 747 would give it an energy of about 32,000 joules/kilograms. But that wouldn't be anywhere near enough to blow up the Borg cube. The Space Shuttle would do a better job. The Shuttle holds the human airspeed record from reentry, when it plunged into Earth's atmosphere at 17,500 mph. This would give the Shuttle an energy density of about 31 million joules/kilograms, still way short of our required 216,237,600 joules. (Incidentally, a typical housefly propelled at Shuttle speed would have the same energy as a .45ACP bullet at point-blank range. Speed is key.)

We'd have to leave our solar system to find objects moving fast enough to even dent the Borg cube. For example, typical debris which is floating free in our Galaxy is moving with an energy density of about 20 billion j/kg. That's enough to blow up the Borg cube! It means that a typical 21-gram African swallow traveling at such galactic speeds would have an energy of 420 million joules, which is almost twice the energy we need. However, one suspects that Species 8472's access to African swallows is limited, as is the means to accelerate them to galactic speeds.

A MATTER OF FLUIDIC SPACE

Let's consider antimatter. Antimatter is made of material opposite in all ways to ordinary matter. The idea of antimatter was first mooted by English physicist Paul Dirac in 1930. The existence of the positron, or anti-electron, was confirmed two years later. The potential energy of antimatter is colossal. As Ensign Garrovick says to Kirk in *The Original Series* episode "Obsession," "Just think, Captain, less than one ounce of antimatter here is more powerful than ten thousand cobalt bombs," though the use of the word antimatter is a lot looser in the franchise than in actual science.

What if Species 8472 built an antimatter missile? Antimatter missiles could be delivered into the very heart of the Borg cube, through the twists and turns of the cube's alcoves and corridors. While it's true that antimatter cannot easily exist in our Galaxy, perhaps fluidic space has far more antimatter than does our space. Species 8472 would no doubt need tons of antimatter for this job, whereas only trillionths of a gram have been isolated in real terrestrial labs.

Why would antimatter make such a monumental missile? Because when antimatter combines explosively with ordinary matter, the result is 100 percent mutual annihilation. Einstein's famous equation $E=mc^2$ tells us that a small amount of mass, "m," converts into an enormous amount of energy, "E," as the "c" in the equation, when squared, is the number 8.99×1016. A mass of antimatter equivalent to a car would produce all of the world's electricity for one year.

Species 8472 could use a form of tractor beam to project a force field that manipulated an antimatter missile into the heart of the Borg cube. Indeed, if they have cunning about them, and they certainly seem to, they could make missiles of anti-Undinium (given the fact that the multiplayer game Star Trek Online offers Species 8472's proper name as Undine). The missiles of anti-Undinium would be made of a super-dense material, with millions if not billions of kilograms per cubic centimeter.

The anti-Undinium missile would be fired into the core of the Borg cube. Species 8472 would certainly have the tech to ensure that this first missile ran down the twists and turns of the cube's corridors, avoiding contact and annihilation with ordinary matter. Next, a missile of regular Undinium would also be sent into the core, at a time cleverly calculated to meet the first missile head-on, at the very core of the cube, doing most damage. The two missiles would meet and annihilate one another and the Borg cube simultaneously.

HOW MUCH WOULD IT COST TO BUILD THE ENTERPRISE?

"What matters is not what they look like now, but what they looked to others at the time that they prevailed . . . There is only one spaceship that's earlier than [the original Enterprise], and that's the flying saucer from 'The Day the Earth Stood Still.' So, what matters here is, what did [the Enterprise] look like at the time it came out (1966) compared with anything that had been imagined before? And when you consider that, that is the most astonishing machine that has ever graced the screen."
> —Neil deGrasse Tyson, Comic-Con (July 16, 2012)

"The Enterprise was the first ever spaceship represented in storytelling that was not designed to go from one place to another; [it was] only designed to explore. It was revolutionary in terms of what we would think space would, and should, be about."
> —Neil deGrasse Tyson, *National Geographic*
> (November 27, 2015)

THE STARSHIP ENTERPRISE

Science fiction like *Star Trek* has often been a home for fantastic and amazing designs. Unfettered by reality, unbound by basic physics, designers and creative artists have been able to bypass the distraction of actual construction and focus instead on imagination and imagery. Little wonder that the iconic works of sci-fi are so fixed in our collective consciousness.

And so it is with spaceships. The web is replete with suggestions about the top ten spaceships of all time: the TARDIS from Doctor Who, the USS *Cygnus* from Disney's 1979 movie *The Black Hole*, Gerry Anderson's classic *Thunderbird 2*, the Vorlon transport in *Babylon 5*, the Millennium Falcon and the TIE Fighters from *Star Wars*, and the Borg cube and the Starship Enterprise from Star Trek.

The silhouette of the famous Federation starship has changed little since *The Original Series*. The ship comprises a saucer section and two nacelles combined with the main body in a design that is set apart from almost all other contemporary sci-fi designs. *Star Trek* designer Matt Jefferies rejected the idea of propulsion, at Gene Roddenberry's request, and tackled the concept of the flying saucer, but with a twist. All future starships kept the same design with minor adjustments: a flatter type for the Enterprise-D, a pointed nose for Voyager, and so on. Passing the test of time and durability, the Enterprise is truly iconic.

BEGINNING THE BUILD

THE RAW ENTERPRISE

So, let's build an Enterprise. Or at least a starship as close as we can possibly get with today's science and engineering. Our ship may not have the jaw-dropping tech of warp drives and teleporters, impulse engines and deflector shields, but we shall do our best. As we don't want to upset any purists, we shall not build a specific Enterprise, but a generic one. Let's aim for an Enterprise around half a mile long, a tenth of a mile high, and with a saucer section of about a fifth of a mile across. These starships are big beasts.

With a structure of this size in mind, the very shell itself is going to cost a pretty penny; how will we estimate the price of the raw materials? Probably the closest current ship we have to the Enterprise sails not in space, but on the seas. After all, we learned earlier in this book that the reason we call them spaceships in the first place harks back to the early days of sailing the open seas and, inspired by the telescope, Kepler declaring that in the far future there would be explorers who would look for new worlds in space, much like medieval sailors discovered new "worlds" on

Earth. Incidentally, talking of big beast ships, some truly mammoth early ships were those associated with Zheng He, a Chinese mariner, explorer, fleet admiral, and court eunuch during China's early Ming dynasty. A fifteenth-century Kirk, Zheng was boss of the expeditionary treasure voyages to Western Asia and Southeast Asia, the Indian subcontinent, and East Africa from 1405 to 1433. And, according to legend, his larger treasure ships used by the commander of the fleet and his deputies were nine-masted, about 417 feet long, 171 feet wide, carried hundreds of sailors on four decks, and were almost twice as long as any wooden ship ever recorded. Comparing his fleet to that of Columbus, Zheng's had 28,000 sailors on 300 ships. Columbus in 1492 had 90 sailors on three ships, the biggest of which was a mere 85 feet long. Zheng's armada even included so-called Equine ships to hold their horses and as many as twenty tankers to carry fresh water.

Now, consider the nuclear-powered aircraft carriers named after WWII US Pacific commander Fleet Admiral Chester W. Nimitz. With an overall length of about a fifth of a mile and full-load displacement of over one hundred thousand tons, it would take around two aircraft carriers worth of material to build our Enterprise, and so the estimated cost of our raw materials is a mere **$13,750,000,000.**

THE ENTERPRISE CONSTRUCTION

What about the construction costs? Here we could take our lead from the way the International Space Station (ISS) is assembled. When complete, the ISS covers an area the same size as a soccer field and weighs 455 tons. Impossible to build on the ground and then launched into space in one go, as there was no rocket big or powerful enough, the ISS was gradually built in space, piece-by-piece, approximately 400 kilometers above the Earth's surface.

Let's assume our Enterprise is also built on Earth, as a set of small but integrated series of modules, which are then assembled in orbit. This is where costs can become pretty cosmic. Using published figures from SpaceX, for its partially reusable heavy-lift launch vehicle the Falcon Heavy, it may be possible to transport our modules into space for the competitive cost of just over $1,000/pound. A pair of aircraft carrier's worth of cargo weighs around 228,000 tons, so at $1,000 per pound, it would cost approximately **$520.4 billion** just to get our Enterprise-shaped space station into orbit. And we haven't yet included labor.

PIMP MY RIDE, MATTERSHIFTER

Our Enterprise sits in space and is now ready to be kitted out with crew quarters, life support systems, computer networks, power generators, and even 5-star luxury options, such as a Ten Forward lounge. To fit our starship all mod cons we shall use a suite of "mattershifters." But what are mattershifters?

Well, in March of 2018, a startup company with alumni from MIT and Yale made a breakthrough. They created a next-generation material that may make it possible to 3D print literally anything out of thin air. The New York–based MatterShift managed to make large-scale carbon nanotube (CNT) membranes that are able to combine and separate single molecules. The company claimed that their tech gave a level of control over the material world never seen before. For example, removing CO_2 from the air and turning it into fuels. Using such tech, it may be possible to produce carbon-zero gasoline, diesel, and jet fuels that are cheaper than fossil fuels.

This is where the mattershifter is crucial to building our Enterprise. Such machines in the future will be able to combine different types of CNT membranes, for example, in machines that can make anything from basic molecular building blocks. And as we are talking about printing matter from thin "air," we can build our Enterprise in space, printing all we need from asteroidal material transported into Earth's orbit. We won't include any extra transportation costs for asteroid material. A class of asteroids known as easily recoverable objects (EROs) was identified by scientists back in 2013, so we can assume asteroid mining will already have evolved from a new space frontier into a run-of-the-mill fact, with asteroids mined using existing rocket science and relatively cheaply.

We can't possibly kit out our huge starship with a single mattershifter, so let's work out the cost of sending a suite of one hundred mattershifters into orbit. Using an estimate of $5,000 per mattershifter (total $500,000), and adding a launch cost of $250,000, we have a total pimp cost for the Enterprise of **$750,000.**

THAT'S ENTERTAINMENT!

We all love the Holodeck. A starship without one wouldn't feel complete, so how much would it cost to build one? Sure, we know that the Holodeck is a fictional device. It's a technical convenience which allows a plot's characters to engage with different virtual reality scenarios. And, from a tale-telling perspective, it enables writers to include a huge variety of other characters and locations, like people and events from Earth's history or alien places and beings, which would otherwise need complex storylines of time travel or dreamscapes.

And yet the role of virtual reality scenarios has become a main feature of our modern lives. The first appearance of a Holodeck came in the *Star Trek: The Animated Series* episode "The Practical Joker," where it was called the recreation room. But it wasn't until the 1988 *Star Trek: The Next Generation* episode "The Big Goodbye" that the Holodeck played a central part of the plot. By the early 1980s, video arcade games had reached their zenith. The arcade video game industry's revenue in the United States was $2.8 billion in 1980, $5 billion in 1981, and $8 billion in 1982. Since the 2010s, the blossoming Asian markets, with games on mobiles and smartphones, have driven growth in the industry even further. By 2018, video games generated $134.90 billion/year of sales globally and became the third-largest sector in the US entertainment market after broadcast and cable TV.

With all this real-life progress, we might not be able to build a Holodeck just yet, but we surely could build a system somewhere between a recreation room and a Holodeck, somewhere crew members could jack into the matrix. Back in late 2018, San Jose holographic display company Light Field Lab and LA graphics company OTOY, who focus on cloud-based high-end graphics, officially announced a partnership whose aim is to make "the Star Trek Holodeck a reality."

Gene Roddenberry founded Roddenberry Entertainment in 1967. The company is currently headed by Gene's son Rod Roddenberry, whose support of this Holodeck-inspired project lends kudos to the concept as compared to many other attempts to market holographic imagery in the past. Rod Roddenberry is impressed by the Light Field Lab and OTOY tech: "The concept of the Holodeck was extremely important to my father as well as the Star Trek Universe. I want to see *Star Trek*'s technologies made real, and for the very first time, now believe that a real Holodeck is no longer limited to science fiction."

That's a good enough endorsement for our Enterprise build, so let's take the Light Field Lab and OTOY tech as our go-to Holodeck. Admittedly estimating more wildly than we would like, let's estimate the cost of our system into orbit and up and running to be a cool **$10,000,000**.

"WAR! WHAT IS IT GOOD FOR?"

Since people who can't kill will always be subject to those who can, and it's a big old Galaxy out there, we'd better weapon up. As photon torpedoes and phaser arrays seem to be pretty standard, let's aim for something like that. The cost of getting three dozen Trident IIs into space (USS *Voyager* was a science ship and even that had thirty-eight photon torpedoes), and pretending they're as good as photon torpedoes, is approximately, give or take half a billion, $6,750,000,000, with a further $250,000,000 for the AN/SEQ-3 Laser Weapon System, which is our approximation for an Enterprise phaser bank, making a weapons total of **$7,000,000,000.**

"HEY SKY, TAKE OFF YOUR HAT, I'M ON MY WAY!"

And finally, so that our Enterprise isn't some kind of latter-day galactic *Mary Celeste*, we need staff. Now, it's no doubt true that staffing costs are not building costs, strictly speaking, and yet it feels as though we wouldn't be doing a proper build job if we didn't think about kitting her out with crew, even if it were a skeleton command crew. The Enterprise D's initial departure from Starbase 74 was carried out with no active crew aboard, using entirely automated procedures. And in the *Star Trek: The Next Generation* episode "Remember Me," the Enterprise D is left with just Dr. Beverly Crusher and Picard on board. But such examples are anomalous.

Let's really splash out on our skeleton command crew and go for a dependable dozen. But at what price? A recent accord between the Russians and NASA put the cost of training a cosmonaut and getting them from the blue into the black at roughly $80 million per person, at 2021 prices. Let's assume our crew are already well trained by Starfleet, and need minimal further training, thus giving us a transport cost of $960,000,000.

No doubt we could staff our starship with bots and droids, as it would certainly cut down on living expenses. But let's keep this human. Between 2010 to 2012, when the United States for a time had more than 100,000 soldiers in Afghanistan, the cost of the war grew to almost $100 billion a year, according to government figures. So, with inflation, let's plug in a cost of $1.5 million per crew member, as we are dealing with officer class recruits heading for space. That adds another $18 million to our transport cost of $960,000,000, giving a running staffing total of $978,000,000.

Unlike droids, our human crew must be fed and cared for. In recent history, NASA awarded a roughly $4 billion contract to private contractors to do the job of conveying cargo to the ISS. Let's use that estimate, giving us a staffing final total cost of **$4,978,000,000.**

WHAT'S THE FINAL DAMAGE, GUVNOR?

Let's look at our running totals. The estimated cost of our raw materials is **$13,750,000,000.** It's **$520.4 billion** to get our Enterprise into orbit; a total pimp cost of **$750,000**; a cool **$10,000,000** for the onboard entertainment; a weapons total of **$7,000,000,000**; and a staffing cost of **$4,978,000,000.**

All these costs give us a grand total of **$546,138,750,000**, or very roughly **$546 billion**. Now, if that sounds a huge amount of investment, just remember that Neta Crawford, chair of the political science department at Boston University, estimated that the long-term cost of the Iraq War for the United States was $1.922 trillion. That's enough to pay for roughly three and a half starships on our agreed budget. And that's without the estimated 1,033,000 excess deaths due to war. Perhaps a better comparison was the giant leap for mankind of the Apollo program, whose total cost was around $152 billion in today's dollars. So, our Enterprise would roughly cost the same as three and a half Apollo programs.

Only one snag, of course. Our Enterprise won't budge a galactic inch, let alone a light-year. Contemporary science is still so far away from warp drives that we can't even make a realistic estimate. The cost will probably be astronomical. Nevertheless, we have a pretty cool weaponized space station, even if it doesn't look like a Moon. . . .

WHAT DOES THE MEASURE OF A MAN SAY ABOUT MACHINE SLAVERY?

"Science must be understood as a social phenomenon, a gutsy, human enterprise, not the work of robots programmed to collect pure information."
— Stephen Jay Gould, *Star Trek: The Next Generation* (1981)

"We pass through this world but once. Few tragedies can be more extensive than the stunting of life, few injustices deeper than the denial of an opportunity to strive or even to hope, by a limit imposed from without, but falsely identified as lying within."
— Stephen Jay Gould, *Star Trek: The Next Generation* (1981)

THE MEASURE OF A MAN

"The Measure of a Man" was the ninth episode of the second season of *Star Trek: The Next Generation*. The episode story, which has themes of slavery and the rights of artificial intelligence, is arguably one of the best-ever tales in the franchise. The episode examines the way in which cyberneticist Commander Bruce Maddox attempts to attack Commander Data's right of self-determination, so he can declare Data mere property of Starfleet and thereby dismantle the android in order to create clones of him. The Enterprise crew are caught up in this struggle. Picard defends Data in a Starfleet court, while Riker is compelled to represent the scientist Maddox.

Indeed, the very soul of the story of "The Measure of a Man" is struggle. As French philosopher and author Albert Camus once wrote, "There is scarcely any passion without struggle." Struggle is at the core of most great tales. It's a spring of suspense and spectacle. And yet, this recognition to some extent puts *The Next Generation* at odds with *The Original Series*. At times it seemed Gene Roddenberry suggested there should be little or no struggle on the Enterprise. In his great future, you could forget the struggles of the past.

During the development of the story, episode writer Melinda M. Snodgrass was called to task by Gene Roddenberry over the apparent conflict between Enterprise crew members. In commentary on the episode, Snodgrass explains how Roddenberry was one of the biggest problems in getting the story to the screen: "Here was this frustration that we couldn't have conflict among the crew, or it was very difficult to get it. Part of that was due to Gene. He once told us that by the twenty-fourth century people were perfect. He said 'my people are perfect; they have no flaws' in one meeting. I was taken aback by it because Gene was a writer. He'd created the show originally, and knew that the essence of drama is conflict—if you don't have conflict, you don't have drama." Snodgrass was even called to California to meet with Roddenberry, who told her "There are no lawyers in the twenty-fourth century, because we don't need them." Eventually, however, the situation resolved and, as Snodgrass said, "Whatever happened, happened above my pay grade, I don't know—but suddenly there were lawyers in the twenty-fourth century!"

CONFLICTING FUTURES

This struggle between Snodgrass and Roddenberry is fascinating: new versus old, *Original Series* philosophy versus *Next Generation* philosophy. What would have Roddenberry's "The Measure of a Man" looked like? Maybe it would have featured Data offering his android parts up for exploitation, giving himself over to the state in the form of Starfleet, so they might develop such droids further. We can assume, if that had been the case, that the episode would have featured little conflict about Data's rights as an AI, which would have called into question the assumed utopian nature of Roddenberry's future vision.

We can read "The Measure of a Man" as a gentle refutation of the kind of future offered by *The Original Series*. The future would not be free of struggle, at least not at an individual level. Roddenberry once wrote that "What *Star Trek* proves, as faulty as individual episodes could be, is that the much-maligned common man and common woman has an enormous hunger for brotherhood." And yet the challenge that faced the writers of *Next Generation* was to create compelling characters and tales, given that so much drama is based on the kind of petty conflict that's supposedly disappeared in the Star Trek Galaxy. The crew's characters would ideally be consistent with the economy and culture of the society in which they swim. A society without need or want. Such future humans would be very different people, perhaps fascinated by the higher things in life, like Picard's fascination with archaeology, or cutting-edge science and the further cultivation of the mind.

LAW AND DEMOCRACY

And yet, surely the crew of the Enterprise would be interested in a workplace drama that pitted the greater good of the Federation against the individual rights of an AI like Data? Interestingly, Melinda M. Snodgrass holds a law degree from the University of New Mexico School of Law, as well as a history degree from the University of New Mexico. Both academic disciplines favor the importance of conflict as a determining factor in the nature of society. The assumption is that, even in the post-scarcity Galaxy of *Star Trek*, humans will sometimes come into conflict. All joking aside, maybe the future will even find a need for lawyers, as they would have a role to play in the democratic process.

This relationship between law and democracy is a long one. Revolutionary democrats like Mahatma Gandhi, Nelson Mandela, and Rosa Luxemburg all studied law. And in ancient Greece, the very ability to make arguments was extensively promoted by the intense political culture of everyday Greek life. Importance was placed on deals in trade and law, where each man represented himself, and judges were chosen at random, and this led to the development of debate to the highest possible level. As a result, the Greek city-states allowed far superior prospects for the typical citizen than did the capital of one of the great palace empires.

CONFLICTING SERIES

The view of the Federation, from *The Original Series* on, was of a society founded upon selfless and idealistic individuals, who are seldom if ever in conflict, and are never in conflict with the greater good. It's surely an idea of society worth striving for and is one that meant much to Roddenberry and to fans of the franchise. (Jonathan Frakes often quotes Roddenberry's words that "in the twenty-fourth century there will be no hunger, there will be no greed, and all the children will know how to read.") "The Measure of a Man" refines this idea somewhat by asking the pertinent question: What happens when the greater good, represented by the state through the Federation, comes up with demands that don't chime with an individual's freedom of choice?

In a rather twisted idea of utopia, if the Federation came calling to clone Data, instead of never asking Data to submit himself to the state in this way, Roddenberry suggests the android would happily give up his individual rights. In "The Measure of a Man," we can see that this statist view of Roddenberry's is represented by the character of the cyberneticist Bruce Maddox. Like other creatives, Maddox leaves more than a little to be desired when it comes to emotional intelligence and, well, reality in general. Maddox is single-mindedly obsessed with the utopian potential for his work: "Consider, every ship in Starfleet with a Data on board. Utilizing its tremendous capabilities, acting as our hands and eyes in dangerous situations. If I am permitted to make this experiment, the horizons for human achievement become boundless." But when Data begs to differ with utopia, Maddox arrogantly refuses to recognize that anyone, least of all a mere robot, has the right to stand in the way of utopian progress: "Rights! Rights! I'm sick to death of hearing about rights! What about my right not to have my life work subverted by blind ignorance?" Maddox is a visionary scientist, a man building for the "year million," incensed by what he sees as the limited vision of those around him, given the pressures of the inspired work he's planning.

ECHOES OF SLAVERY

The main thrust of "The Measure of a Man," of course, is its compelling account of how we might define life in the future, when the needs of artificial life achieve some kind of parity with so-called 'natural' life, and how the state might deal with the rights of artificial persons. As science has yet to develop a real-life Data, the debate is understandably abstract. And yet it has far-reaching consequences in time, as Picard argues in one of his monologues: "And the decision you reach here today will determine how we will regard this creation of our genius. It will reveal the kind of a people we are, what he is destined to be. It will reach far beyond this courtroom and this one android. It could significantly redefine the boundaries of personal liberty and freedom, expanding them for some, savagely curtailing them for others. Are you prepared to condemn him and all who come after him to servitude and slavery? Your Honor, Starfleet was founded to seek out new life. Well, there it sits. Waiting. You wanted a chance to make law. Well, here it is. Make a good one."

Is this not a superior declaration of the philosophy of the franchise than a vulgar and rigid insistence on utopia? Picard is suggesting that, if humans have already found perfection, what more is there to discover? Picard's perspective is ably supported by Guinan, an alien character who is several hundred years old and is noted for her warmth and folk wisdom, who suggests that the trial of Data shares some similarities to the treatment of slaves who were denied personhood in the service of more "important" powers: "Well, consider that in the history of many worlds there have always been disposable creatures. They do the dirty work. They do the work that no one else wants to do because it's too difficult, or too hazardous. And an army of Datas, all disposable, you don't have to think about their welfare, you don't think about how they feel. Whole generations of disposable people."

Picard and Guinan's future-proof perspective on AI chimes well with the 2016 dramatization of Michael Crichton's *Westworld*, the story about adult theme-park androids rising up against their creators. HBO's retelling of the tale is fascinating. *Westworld* remains a holiday destination for privileged dilettantes, with the tourists mostly abusing their robotic hosts for their own cheap ends. Like Maddox, these prosperous humans seldom stop to think about the moral consequences of transgressing the rights of a conscious humanoid. As with Picard's attitude to Data's position within the Enterprise and wider Federation, the AI in Westworld is so thoroughly incorporated into contemporary life that it's a given that the evolution of AI consciousness means new needs and rights for the humanoids. It's an inevitable conclusion that the theme-park creators must tackle the tricky question of whether the androids should be granted a soul. Surely, it would be ethically wrong to deny the humanoids their full potential. It would be a human moral duty to pass on the torch of evolution from apes to us to whatever comes next.

And so "The Measure of a Man" was a huge episode for *The Next Generation*. The story showed that the series warranted comparison with *The Original Series*, that it didn't shy away from big ideas. "The Measure of a Man" asks some important questions: once we accept that humans have arrived in a new place technologically, such as the advent of a humanoid like Data, how do we deal with it? Do the beings we create deserve a chance to live their lives independently of human meddling from the likes of Maddox? What moral guidelines must be drawn in the use of developing tech, and what is the role of such tech in our lives?

AS COMMANDER RIKER ASKS, "WHAT EXACTLY IS A DYSON SPHERE?"

"This idea can be traced back to the work of Freeman Dyson whose 1959 scientific paper "Search for Artificial Stellar Sources of Infra-Red Radiation" appeared in *Nature* and proposed orbiting structures, either spheres or swarms, designed to catch and collect all the energy radiating from a star . . . This process, Dyson suggested, would alter the wavelength of the energy emitted by the star and this change would allow astronomers to discover any spheres or swarms that already existed. Here, Dyson reasoned, may lie proof of alien life."

—Mark Brake, *FutureWorld* (2008)

RELICS

In the *Next Generation* episode "Relics," the Enterprise responds to a nearby distress call and discovers a Dyson Sphere. The origin of the call was the USS *Jenolan*, which has crashed on the Sphere's outer shell. The Jenolan, a Federation transport ship that has been missing for seventy-five years, has an interesting passenger in none other than former Starfleet officer Captain Montgomery Scott, a.k.a. "Scotty" from *The Original Series*. In the episode, Riker asks after the nature of a Dyson Sphere. Picard replies that the twentieth-century physicist Freeman Dyson postulated the theory that an enormous hollow sphere could be constructed around a star, which would have the advantage of harnessing all the star's radiant energy. A population living on the interior surface would have virtually inexhaustible sources of power.

DYSON'S THOUGHT EXPERIMENTS

The eponymous sphere was the result of Dyson's thought experiments about rebuilding the solar system. Dyson had first theorized that technological progress would happen very rapidly once a civilization had entered the technological age. Compared with the vast eons of astronomical and geological time, such technological progress is swift indeed. And yet one crucial factor that could restrict the speed of progress would be the limited supply of energy and matter.

Dyson's math went something like this. In Dyson's contemporary world, the material resources available to humans are limited to the Earth's biosphere which comes in at roughly 10-8 of the mass of the Earth. Meanwhile, the energy needed by contemporary society in Dyson's day (though he's still alive as I write this!) is about equal to the energy liberated from burning between 1 and 2 billion tons of anthracite each year. As a result, the Earth's resources of fossil fuels would be exhausted in a few centuries. Dyson discovered that, if he modestly projected such figures into the future, the rate of manufacture in 2,500 years will increase by 10 billion times or be about 0.01 percent of the entire luminosity of the Sun! And so, Dyson was forced to consider the question: If our progress continues, and we head for energy demands that measure in cosmic proportions, when will we exhaust the energy resources of Earth?

DYSON'S SPHERE

Dyson came to a rather startling conclusion. He found that, even if he included nonfossil fuel energies, such as wind, nuclear, and solar, the energy resources of Earth are no way enough to satisfy future human civilizations, assuming growth and progress continue. So, Dyson put his mind to the material resources available off-Earth. He decided that a highly technical human civilization of the future could make use of the resources in the wider solar system. In particular, the potential source of material that is held in the Jovian planets.

Dyson decided the mass of Jupiter could be used to build an immense shell around the Sun. Yes, you read that right. His famous project is of *Star Trek* proportions. Dyson calculated that Jupiter could be vaporized and used to make a sphere with a radius of 93 million miles. This is no arbitrary choice. Ninety-three million miles is 1 Astronomical Unit, or AU, the distance from the Earth to the Sun. Dyson's Sphere is some build! How thick would the shell of the Sphere be? Assuming the material used for the hollow Sphere is the mass of Jupiter, most of that mass would be used to make up the immense surface area of the Sphere. And that would mean the thickness of the shell itself would be only a few meters.

What would it be like, living on the inside surface of such a Sphere? Well, practically speaking, humans are already two-dimensional animals. We live on the outer surface of the sphere of planet Earth, so it might be perfectly feasible, in the millennia to come, for humans to create an artificial biosphere on the inner surface of a Dyson Sphere. (Since Dyson, such artificial biospheres have become popular in science fiction. Think of the built environments in Arthur C. Clarke's 1973 book *Rendezvous with Rama*, with its cylindrical alien starship, the Halo video game franchise, with its ring-shaped megastructures, and the 2013 movie *Elysium*, with its gigantic space habitat in Earth's orbit.)

After completing this gargantuan civil engineering project, human society would be able to use the total energy output of the Sun. Every solar photon could be utilized, falling as it would on the inner surface of Dyson's Sphere. Furthermore, as the total surface area of the Sphere is around one billion times the surface area of the Earth, our new living space could support a huge future population. (Let's not worry how such a Sphere might be built, how it might rotate, or how we might prevent its human inhabitants from falling into the Sun!) The main point is that Dyson expects any advanced intelligent species to reach the stage of occupying an artificial biosphere, which completely surrounds the parent star.

SEARCHING FOR SPHERES

How will an advanced alien civilization living on the inner surface of a Dyson Sphere appear from outside? In the words of Dyson's 1960 academic paper, "Search for Artificial Stellar Sources of Infrared Radiation," "If the foregoing argument is accepted, then the search for extraterrestrial intelligent beings should not be confined to the neighborhood of visible stars. The most likely habitat for such beings would be a dark object, having a size comparable with the Earth's orbit, and a surface temperature of 200 to 300 K. Such a dark object would be radiating as copiously as the star which is hidden inside it, but the radiation would be in the far infrared, around 10 microns wavelength." If the civilization did not let out the infrared energy, the radiation produced by the star would build up inside the Sphere and result in disastrously high temperatures.

So, an advanced alien civilization living inside a Dyson Sphere would be a powerful source of infrared radiation. This is very convenient for humans looking for signs of extraterrestrial intelligence. The atmosphere of the Earth is transparent to radiation in regions that accommodate the search for such infrared sources. Humans already have the technology that would enable the detection of Dyson Spheres over great distances. Though, naturally, we would have to research ways of detecting the difference between artificial infrared sources such as the Spheres and natural objects like protostars, those contracting gas masses in the formation of a star that emit infrared radiation with the same intensity.

A sky chart could be compiled, with potential Dyson Spheres mapped out in the local quadrant of our Milky Way. Each source could then be investigated by other techniques for signs of an intelligent species. And maybe this is why Dyson's idea made a starring appearance in the *Next Generation* episode "Relics." The Dyson Sphere is a good example of how the culture of an intelligent alien society might change a planetary system so much that it would be detectable over interstellar distances, or if you just happened to be passing nearby in a twenty-fourth-century starship.

A number of scientific research projects, including SETI (Search for Extraterrestrial Intelligence) and Fermilab, using data from the Infrared Astronomical Satellite (IRAS), have used strategies associated with Dyson's ideas. Fermilab found seventeen potential "ambiguous" candidates, of which four have been called "amusing but still questionable." But the main point is this. You'd have thought that such a fantastical idea as the Dyson Sphere would remain the plaything of science fiction like *Star Trek*. And yet the prospect of these megastructures is so captivating that scientists have actually spent time looking for the Spheres. It's a testament to the power of ideas.

APOCALYPSE COW: IS THE *STAR TREK* REPLICATOR THE FUTURE OF FOOD?

"It just makes you think, how did people know what to want? I mean, if it's anything you like, any time, it's like science fiction, where they have a machine that just makes stuff. It does your head in. Press a button, and it's roast beef, pheasant mole, chickpea fritters in yogurt dressing, aioli, prawn curry, mango soufflé, duck blood stir fry, consommé, you know, where does it all end? I mean, the idea is amazing, everything all the time, I get it, and yet, it's weird and wrong too."

—John Lanchester, *The Wall* (2019)

"Replicators are the worst thing ever. Destroys storytelling all the time. They mean there's no value to anything. Nothing has value in the Universe if you can just replicate everything, so all that goes away. Nothing is unique; if you break something, you can just make another one. If something breaks on the ship, it's 'Oh, no big deal, Geordi can just go down to engineering and make another doozywhatsit.' Or they go to a planet and that planet needed something: 'Oh, hey, let's make them what they need!' [The writers' room] just hated it and tried to forget about it as much as possible."

—Ronald D. Moore, *Bleeding Cool* (2017)

THE *STAR TREK* REPLICATOR

In the Star Trek Galaxy, a replicator is a machine that makes or recycles stuff. We first meet these machines in *The Original Series* where they were simply used to synthesize meals on demand. But the actual term "replicator" wasn't used until *The Next Generation*, where it is described as a twenty-fourth-century advancement from the twenty-third-century "food synthesizer."

According to Mieke Schüller in his 2005 book *Star Trek - The Americanization of Space*, "The so-called 'replicators' can reconstitute matter and produce everything that is needed out of pure energy, no matter whether food, medicaments, or spare parts are required." Schüller's theory is based on an idea in physics known as "the equivalence of mass and energy." This equivalence, first correctly described by Albert Einstein, was said by the man himself to be "the most important upshot of the special theory of relativity" and is neatly summed up in science's most famous equation, $E = mc^2$.

While most people focus on the destructive power of converting a small amount of mass into the huge amount of destructive energy associated with, for example, hydrogen bombs (remember that the "c" in Einstein's equation is the speed of light), Schüller reminds us that the equivalence also works the other way. Energy can be used to make mass out of "nothing." As Einstein showed, given that energy and matter are just aspects of the same thing, it follows that matter can come from pure energy. Indeed, that's where matter in the Universe originally comes from—physicists believe matter in the cosmos is simply frozen energy.

In the fictional physics of *Star Trek*, a replicator can make any inanimate matter, as long as the given material structure is known and on data file. What it cannot manufacture, at least in the case of Federation replicators, is antimatter (remember that when it comes into contact with ordinary matter, antimatter and matter mutually annihilate one another, so not a good idea), dilithium, and the fictional currency, latinum. Nor can Federation replicators create living things. In the non-canon work *Star Trek: The Next Generation Technical Manual* it is suggested that even though replicators share similar tech with transporters, the technical spec lacks the "resolution" to create living tissue. There are, however, other

replicators within the fictional physics of *Star Trek*, like those used by the race of aliens in *The Next Generation* episode "Allegiance" to create living things, including the brain's many trillions of dendric neural connections where memory is stored. But what relevance does the *Star Trek* replicator have to our own future?

APOCALYPSE COW

We could be on the cusp of the biggest economic revolution for two hundred years. New tech will soon mean that most of our food will come neither from fauna nor flora, but from unicellular life. After ten thousand years of using farming to feed humanity, all food production, save that of fruit and veg, will be supplanted by "ferming," the precision fermentation of brewing microbes. And that means the multiplication in factories of certain microorganisms to make particular products.

Humankind is refining its own version of the *Star Trek* replicator. For example, in a lab on the outskirts of Helsinki, technicians are turning water into food. Metal tanks replete with a yellow churning froth, a primordial soup of bacteria, are turning the froth into a rich yellow flour. The bacteria are drawn from the soil and multiplied in the lab, using the hydrogen taken from water as the energy source. The company is called Solar Foods and such flours will soon be the feedstock for almost everything. In their raw state, the flours can be substituted for the fillers now used in a huge number of food products. After modification, the bacteria can be used to create the particular proteins required for lab-grown milk and eggs and, yes, even meat. Further tweaks will produce the long-chain omega-3 fatty acids needed to make lab-grown fish. The carbohydrates left over when fats and proteins have been removed can be used to make everything from pasta flour to potato chips.

Solar Foods is not the only company revolutionizing food production along replicator lines. Scientists at Nestlé are working on replicator-like tech to create food tailored to a person's nutritional needs. Physicists at Imperial College, London, have found how to make matter from light in the same manner suggested above. And BeeHex, an Ohio-based company, received a grant from NASA and are building food-printing robots for eventual public use.

FORGET THE FUTURE?

The *Star Trek* replicator reconstitutes matter out of pure energy. The hydrogen energy pathway used by Solar Foods is ten times more efficient than photosynthesis, nature's way of using carbon dioxide from the air, water from the ground, and energy from Sunlight to make plant food. And as the Solar Foods foodstuff will be brewed in industrial vats, the land efficiency is about twenty thousand times greater.

All humans on Earth could be well fed using only a tiny fraction of the planet's surface. Not only that, but if the water used in this process is much less than what is used by conventional farming and is electrolyzed with the Sun's power, the best places to build these plants will be deserts. How's that for efficient land use?

Current food production methods are ripping the living world apart. And replicator tech could help us avoid catastrophe in the near future. For example, climate catastrophe could cause what's called "multiple breadbasket failures," through concomitant heatwaves and other phenomena. The UN predicts that, by the year 2050, feeding the world's growing population will need a 20 percent growth in the global water use by farming. And yet water consumption is already at breaking point in many places: aquifers are drying up, rivers never reach the sea, and the glaciers that water half of Asia's peoples are in rapid retreat. A world soil catastrophe threatens the very basis of our subsistence. Huge swathes of land have lost their fertility through erosion and pollution. The supply of phosphates, vital for farming, are shrinking fast. Insect-ageddon threatens a cataclysmic pollination meltdown.

Forget the future, it's hard to see how farming can even feed us until 2050. Let alone until the twenty-fourth century. Farming and fishing are, by a huge distance, the greatest cause of extinction and loss of the diversity and abundance of wildlife. Farming is a main cause of climate collapse, the primary cause of river pollution, and a considerable source of air pollution. Over huge swathes of the planet's surface, farming has supplanted natural ecosystems with crude human food chains. And fishing is the cause of ecological collapse in global seas. So, eating has become a moral minefield. Almost everything we eat has an environmental cost.

FREE THE LAND, AVOID CATASTROPHE

It's time for the *Star Trek* replicator. Replicator-like tech can create farm-free food. And farm-free food can save both people and the planet. Farm-free food will enable enormous areas of land and sea to be handed back to wilderness and nature, which will then trigger carbon drawdown on a titanic scale. Replicator-like farm-free food will also mark the end of the exploitation of animals and signal a stop to the majority of deforestation, pesticides, and trawling. And, if done right, farm-free food could enable affordable and abundant food for everyone.

Research suggests that proteins from replicator-like tech will be around ten times cheaper than animal protein in ten years or so. As these new industrial processes use tiny areas of land, with a hugely diminished need for water and nutrients, farm-free food represents the best opportunity for regreening the world in human history. And food will be healthier as well as cheaper. As farm-free foods are made out of simple ingredients, and not broken down from complex ones, harmful constituents like hard fats, allergens, and other unhealthy components can be filtered out. Meat will be grown on collagen scaffolds rather than on animals. And food in general is likely to be less expensive, better, and far less detrimental to our living planet. Farm-free food will be more reliable and stable and can be grown anywhere, even in places without farmland.

Farm-free food could be crucial to ending world hunger and offers hope where there was none before. We will soon be able to feed the world without devouring it. So much potential has come out of an idea as simple as the *Star Trek* replicator!

PART IV
MONSTER

HOW DOES "THE CHASE" EXPLAIN HUMAN EVOLUTION IN THE *STAR TREK* GALAXY?

"Transport, by imagination, a man from our planet onto Saturn, his lungs will presently be rent by an atmosphere too rarefied for his mode of being, his members will be frozen with the intensity of the cold; he will perish for want of finding elements analogous to his actual existence; transport another onto Mercury, the excess heat will quickly destroy him. Thus Man, the same as everything else that exists on our planet, as well as in all others, may be regarded as in a state of continual vicissitude; thus, the last term of the existence of Man is, to us, as unknown, as indistinct, as the first; there is, therefore, no contradiction in the belief, that the species vary incessantly; and it is impossible to know what he will become, as to know what he has been."
 —Baron D'Holbach, *The System of Nature* (1770)

Picard: "It's four billion years old. A computer program from a highly advanced civilization, and it's hidden in the very fabric of life itself."
 —Joe Menosky and Ronald D. Moore, *Star Trek: The Next Generation*, "The Chase" (1993)

THE CHASE

In the *Star Trek: The Next Generation* episode "The Chase," we are reminded that Picard is a keen student of archaeology. And he is visited by his former mentor, Professor Richard Galen, who tells Picard that he has come across something in his travels which could be the greatest discovery of their time. (As Galen puts it, "I made a discovery so profound in its implications that silence seemed the wisest course. This work has occupied my every waking thought, it's intruded upon my dreams, it's become my life. When finished and I announce my findings, it will be heard halfway across the Galaxy.") Naturally, Picard is intrigued. Eventually, he discovers that there is an embedded genetic pattern, which is constant throughout many different sentient species, and which was left by an early ancestor race (whom we shall call "the Founders") that predates all other known civilizations in the Star Trek Galaxy.

This genetic pattern explains why so many sentient races in the Galaxy are humanoid. In "The Chase," a recording of a female Founder is ultimately discovered. She explains that it is her race who are responsible for the presence of life in the Alpha Quadrant. It seems that, when the Founders first explored the Alpha and Beta Quadrants, there had been no other humanoid life-forms, so they seeded numerous planets with their DNA to create a legacy of their existence after their race had passed. The female Founder ends her message by declaring her hope that the knowledge of a common origin will help produce peace in the Galaxy.

First, hats off to *Star Trek*. Science fiction has a bad habit of using humanoid aliens. Whereas in reality, of course, humanoid aliens are a very improbable species. Anyone who's studied exobiology and evolution knows that aliens are very unlikely to look like humans. We know the real reason science fiction uses humanoid aliens. When portraying human drama on screen it's very hard to relate to a sentient orange gas called Dave. Or a green lizard called Blanche.

But, in dealing with the improbability of humanoid aliens, at least "The Chase" tries to explain that humanoids in the Galaxy were genetically preprogrammed billions of years back. All the aliens look the same because of the genetic engineering of the Founders. Possibly. Let's be honest, this is more than you ever get from *Star Wars*, where there even appears to be, unfathomably, humanoid aliens a long, long time ago, in a Galaxy far, far away.

PANSPERMIA

This plot in "The Chase" is an example of the scientific theory of panspermia. This theory has it that seeds or spores can spread life through the Galaxy, or indeed the wider Universe. The theory also holds that very small organisms, or the biological precursors of life, can be found in the Galaxy and that they created life on Earth and other planets. What appears in "The Chase" is a kind of directed panspermia, a guided evolution.

You may recall something similar to "The Chase" in film director Ridley Scott's 2012 movie, *Prometheus*. At the start of the film, we see a spacecraft depart a planet, leaving behind a statuesque humanoid alien, who drinks an iridescent liquid, causing his body to dissolve. As his remains cascade into a waterfall, his alien DNA falls apart and recombines, hastening the evolutionary process. Later in the movie some archeologists discover that the alien belongs to a race dubbed "the Engineers" rather than the Founders, who are thought to be the "creators" of the human race, and who had visited and seeded the Earth during its primordial stage.

Perhaps the most famous case of guided evolution is Arthur C. Clarke and Stanley Kubrick's seminal movie *2001: A Space Odyssey*. Kubrick claimed his movie provided a "scientific definition of God." There was little filmic drama in Darwinian evolution, just the slow, solid state of inexorable change, so Kubrick and Clarke conjured up a fictional form of "punctuated equilibrium." The movie augments the usual driving force of evolution, long periods of steady change, with the episodic guiding hand of superior alien beings. It is a story of the effective creation and resurrection of humanity. The agency that drives the guided evolution of the human race is an elusive alien race in the shape of the famous black monolith artifacts.

PANSPERMIA ALL-SORTS

Directed panspermia isn't the only flavor. First, panspermia is either interstellar (between the stars) or interplanetary (between planets in the same star system). The transport mechanism of the seeds or spores can include comets, the radiation pressure of stars, and what's known as litho-panspermia, where microorganisms are embedded in rocks.

The idea of radio-panspermia was first proposed in 1903 by Swedish scientist Svante Arrhenius, a distant ancestor of climate change warrior Greta Thunberg. In his article "The Distribution of Life in Space," Arrhenius suggested that microscopic seeds of life could be propagated through space, driven by the radiation pressure from stars, though the mechanism only holds for tiny particles such as single bacterial spores. The concept of litho-panspermia, the transfer of organisms in rocks from one planet to another, is rather speculative and would only really hold through interplanetary space. For example, the interplanetary transfer of inorganic material is well known, as witnessed by the Martian meteorites found on Earth.

One astronomy professor, Thomas Gold, even suggested an accidental panspermia. In such scenarios, Gold envisaged a kind of "cosmic garbage," where life on Earth was the result of an ancient and accidental dumping of a pile of waste products on our planet by alien beings. This idea of cosmic garbage features in the book *Roadside Picnic* by prominent Soviet science fiction writers Arkady and Boris Strugatsky. *Roadside Picnic* was translated into English in 1977 and filmed under the name *Stalker* in 1979 by Andrei Tarkovsky, the Russian film director who also created the acclaimed cinematic dramatization of *Solaris* in 1972.

Roadside Picnic is an alien visitation tale with a difference. The story is set in a post-visitation world, planet Earth. There are now six mysterious zones, regions of our globe that have been touched in some way by an alien visitation, some ten years past. The alien visitors were never seen. But people local to the zones reported loud explosions that blinded some and caused others to catch a kind of plague. Though the visit is thought to have been brief, about twelve to twenty-four hours, the half dozen zones are full of strange events. The cosmic trash left behind by the alien "picnic" means that the zones become contaminated with fatal phenomena. The zones are littered with mysterious objects, or artifacts, whose various properties and original intent is so incomprehensible and advanced it might as well be supernatural.

DIRECTED PANSPERMIA

Let's look at the plot of "The Chase" in a little more detail to work out how the original Founders may have seeded their fictional Galaxy. The tricorder projection explains that the original Founder civilization existed alone in the Galaxy for billions of years. Time enough to develop an advanced and technically sophisticated species. Eventually, they decided to spread their humanoid Founder genome to other planetary systems. The aim was to establish a rich community of humanoids, one that could integrate with alien cultures, one that these first beings had mysteriously failed to do, for all their sophistication.

How would the humanoid seeding in "The Chase" have been done? As we have seen, directed panspermia is all about the deliberate transport of life in space. Francis Crick, decoder of the helical structure of DNA and Nobel Prize winner, suggested that life may have been purposely spread by an advanced extraterrestrial civilization. Directed panspermia has also been identified as a way of securing and expanding human life in the cosmos, which also seems to be the rationale driving the Founders in "The Chase." Pan-biotic programs would seek to sow seeds in new planetary systems nearby and clusters of new stars in interstellar clouds. The target planets might be those where local life would not yet have formed, so that the directed seeding would not interfere with local life. Note that in "The Chase," the unexplored star system Ruah Four is visited and found to be "a class-M planet. Sixty-seven percent of its surface is covered with water. Its landmass contains multiple animal species, including a genus of proto-hominids," and that Indri VIII is "L-class . . . covered by deciduous vegetation, unexplored, with no apparent evidence of civilizations, either present or past. The planet possesses no animal life whatsoever."

HUMANOID REPLICATOR

Seeking to propagate the basic patterns of terrestrial organic life is one thing. Ending up with humanoid aliens derived from the Founders is quite another. As we have seen, we humans on Earth arose as the result of the fortuitous and contingent outcome of thousands of linked events. Any one of those events could occur very differently on other planets, in other climates and environments. Even if the original Founders chose very similar planetary habitats to their own, it's quite easy to see how an intended future history could be sent on an alternative pathway, especially one ending up at extinction.

After all, for as long as biota has been evolving on Earth, species have been going extinct. Scientists have estimated that over 99.9 percent of all species that ever lived on our planet are now extinct, and that the average lifespan of a species is between one and ten million years. Why should evolution be any different on any other world? Earth scientists have learned enough about evolution to predict what aspect of life elsewhere would be common with life on our planet. And they have decided that the most common factor would most likely be replication. At home, all life evolves by the differential survival of duplicating entities, which we may call replicators. On Earth, the replicator that does the job is known to us as DNA.

DNA, or deoxyribonucleic acid, exists simply to make more DNA. There's nearly two meters of DNA squeezed into almost every human cell. Each length of DNA has about 3.2 billion letters of coding. That's enough to enable $10^{3,480,000,000}$ potential combinations. That's an enormous number of possibilities. Imagine the biology of that female Founder on "The Chase," since we can assume her biology works something like ours. Her body would be host to around ten thousand trillion cells. Almost every one of them containing two yards chockablock with DNA. If her DNA were all spun into a single thread, it would make a solitary strand long enough to travel a light-second and back many times over. Founders, like humans, would have had as much as twelve and a half million miles of DNA scrunched up inside.

ENGINEERING HUMANOIDS

So, humanoids are vehicles for DNA. From the Founders on, we would have been host to DNA parasites—our genes. In many ways, the actual function of life, that which would be played out in the planetary habitats of humanoid worlds, is simply DNA survival. But DNA is not floating free. It's locked up in living beings. And though we can't live without it, DNA itself is not alive. Truth is, DNA is among the most chemically inert molecules on Earth. And maybe on those other humanoid planets too.

Curious that something so lifeless should be at the very core of galactic life itself.

Given that all humanoids are descended from the Founders, and that terrestrial humans are DNA-based, we must conclude that DNA is the common replicator in all sentient humanoid species in the Star Trek Galaxy. To date, the evolution of humans, like all other species, has been subject to the law of natural selection. But, with our unraveling of the human genome and the development of genetic engineering, a time will soon come when our future is decided not by selection but by human agency. How did the Founders pull such an evolutionary trick, short of leaving a suitable Adam and Eve on each home planet? How did they engineer against extinction, given DNA's incredible potential for change that's coded within its structure? You will recall that in "The Chase," the Founders had created a design that was "part of an algorithm, coded at the molecular level," and that implanted DNA fragments had "been part of every DNA strand on Earth since life began there, and the other fragments are just as old. Someone must have written this program over four billion years ago. So, four billion years ago someone scattered this genetic material into the primordial soup of at least nineteen different planets across the Galaxy."

It seems as if the Founders had been playing the long game with evolution. As the female Founder says in the final act of "The Chase," "Our scientists seeded the primordial oceans of many worlds, where life was in its infancy. The seed codes directed your evolution toward a physical form resembling ours. This body you see before you, which is, of course, shaped as yours is shaped, for you are the end result." How might they have avoided catastrophe for humanoid forms on the seeded planets, assuming that all seeds took root?

Consider the biological thriller *Darwin's Radio*, an eye-opening novel by American sci-fi writer Greg Bear in 1999, six years after the original air date of "The Chase." In Bear's novel, a new form of retrovirus "plague" has emerged. It controls human evolution by swiftly evolving the next generation while it is still in utero, leading to speciation. The story tells the tale of this plague along with the panicked reaction of the public and the American government to the retroviral disease. Scientists soon discover that, built into the human genome, are non-coding sequences of DNA known in the novel as introns. Certain sections of so-called junk-DNA, which are remnants of prehistoric retroviruses, have been triggered and are translating various large protein complexes. The activation of the plague retrovirus, and its associated speciation, appears to be controlled by a complex genetic network that recognizes a need for modification in the face of environmental pressure. In short, it's a genetic human adaptive response to evolutionary change. And the disease, or rather, gene activation, leaves behind a fertilized egg with 52 chromosomes, rather than the typical 46 characteristic of Homo sapiens.

So, here's one solution. The Founders may have designed an algorithm just like the retrovirus plague in *Darwin's Radio*, an engineered genetic human adaptive response to evolutionary change, one which meant the humanoids in the Galaxy abided. Mind you, the Founders would have also had to engineer an adaptive response to contingent outcomes of thousands of possible evolutionary events, events that occurred before one of the myriad branches of seeded life actually became humanoid. It's just as well the Founders were a technically sophisticated species alone in the Galaxy for billions of years. They surely needed enough time to develop their science way beyond our current understanding and, as science fiction writer Arthur C. Clarke famously wrote in his 1962 book *Profiles of the Future*, "any sufficiently advanced technology is indistinguishable from magic."

FARCICAL EPILOGUE

Perhaps the biggest scientific surprise in "The Chase" is the reaction of the Romulans, Klingons, and Cardassians to the secret of life in the Galaxy. After the female Founder has revealed the origin of humanoid life, most parties seem disgusted at the thought of a common progenitor. But why the shock and surprise? Surely the different flavors of humanoid alien would have had at least some hint of familial relations when they first established contact with one another. Okay, sure, there are superficial differences, such as the scrotal forehead of the Klingons, the slick Beatle-wig of the Romulans, and the ridged, crocodilian skin of the Cardassians. But, guys!

Take first contact between humans and Vulcans, for example. It occurred on April 5, 2063. That night, the Vulcan survey ship the T'Plana-Hath landed in Bozeman, Montana, after tracking the warp signature of Zefram Cochrane's Phoenix, the spacecraft that marked Earth's first successful attempt at achieving warp drive. Merely minutes later, a robed Vulcan, displaying the now familiar split-fingered Vulcan greeting, made the acquaintance of Cochrane and essentially welcomed him and all humans to the galactic club. It beggars belief, to say the least, that it would not occur to Cochrane, a very notable Earth scientist, on first seeing the Vulcan, "Hey, these guys look just like us. I wonder what their genetic makeup is?" After all, human-Vulcan first contact is only 110 years after Crick and Watson determined the double-helix structure of DNA, the molecule containing human genes.

Another historical inconsistency that "The Chase" glosses over is the fact that, by the mid-twentieth century, Vulcans were actively observing Earth and the civilizations that inhabited it. The Vulcan survey ships were making routine flybys through our solar system to keep an eye on how humans were developing. Despite the fact that the Vulcan watchers were limiting their observations to remote studies from high orbit, surely they would have noticed not just the similarity of human and Vulcan physiological form, but also the discovery of DNA? (We know from the *Star Trek: Enterprise* episode "The Forge" that DNA samples were collected from all Vulcan children at birth and kept on file by the Vulcan government.) Scientific inconsistencies notwithstanding, "The Chase" is a brave attempt to use the theory of directed panspermia to explain the humanoids in the Star Trek Galaxy.

WHAT DOES *STAR TREK* HAVE TO SAY ABOUT THE POLITICAL SCIENCE OF WAR?

Come you masters of war
You that build all the guns
You that build the death planes
You that build the big bombs
You that hide behind walls
You that hide behind desks. . . .
I think you will find
When your death takes its toll
All the money you made
Will never buy back your soul
—Bob Dylan, "Masters of War" (1963)

"The strength of a civilization is not measured by its ability to fight wars, but rather by its ability to prevent them."
—Gene Roddenberry, *Earth: Final Conflict,* "Scorched Earth" (2000)

"It is the responsibility of intellectuals to speak the truth and expose lies."
—Noam Chomsky, *The New York Review of Books* (1967)

A BRIGHT FUTURE FOR A MONSTROUS WORLD

Star Trek: The Original Series was created and broadcast during one of the most turbulent times in the history of the United States. For almost three years when the series was first broadcast, between September 1966 and June 1969, American society met a number of crises. In addition to a monstrous war, there was also profound civil strife in the nation's cities, burgeoning crime and economic crisis, student protests and rebellions, and huge potential changes to traditional cultural values and gender roles.

In the face of all this, *Star Trek* presented a bright future for the planet, one where Earth has become an affluent, tolerant world, with no war and little civil conflict. It was a future in which the starship USS *Enterprise* was emblematic of a mature and sophisticated society, capable of holding liberal values and taking them into the farthest reaches of the fictional Galaxy.

In the same month that the first ever *Star Trek* episode was broadcast, September 1966, United Nations Secretary-General U Thant said that he would not seek reelection due to the failure of UN attempts to end the Vietnam War. "Today it seems to me, as it has seemed for many months, that the pressure of events is remorselessly leading toward a major war . . . In my view, the tragic error is being repeated of relying on force and military means in a deceptive pursuit of peace." That same month, the US Department of Defense declared what would be the largest draft call of the Vietnam War, calling for almost fifty thousand men to be drafted into military service for the month of October, the highest number since the Korean War.

THE UNREALITY OF WAR

It was the looming prospect of war that was utmost in the mind of many Americans. The conflict in Vietnam looked as though it might pull the nation apart. This complex and controversial war, which spanned two decades and involved almost twenty nations, greatly impacted American culture. Just think of the great movies it inspired: *Platoon, Full Metal Jacket, The Deer Hunter*, and *Apocalypse Now*, to name just a few. But *Star Trek* got there before the films. It was one of the very first drama series to confront the Vietnam War. As a cultural influence on *The Original Series*, the war was an omnipresent factor in the development of the series bible. The utopian future presented by the franchise in those early days suggested an alternative and better world to the actual world of viewers in the America of the 1960s.

Other drama series were scared of losing viewers or advertisers, but sci-fi can spin a story in an entirely different way. Consider the case of American writer Kurt Vonnegut. Vonnegut was left physically unscathed after the wanton British firebombing of the city of Dresden in 1945. Witness to the senseless slaughter of well over 100,000 people, Vonnegut spent a quarter of a century coming to terms with such outrageous and bloody carnage. It was the largest massacre in European history. To realize the full horror of his experience, Vonnegut turned to science fiction. He finally faced his Dresden demons in his wonderfully creative antiwar novel of the late sixties, *Slaughterhouse Five*, named after the underground meatpacking cellar that saved his skin.

With characteristically black humor, Vonnegut paid tribute to the role of science fiction in *Slaughterhouse Five*. The main character, recently witness to the firebombing of Dresden, considered that people were "trying to reinvent themselves and their Universe. Science fiction was a big help." Vonnegut used science fiction to confront the growing horrors of the twentieth century. Perhaps if his readers could stand the unreality of science fiction, they could face a little bit more reality after reading *Slaughterhouse Five* than they could before they read it.

As Matthew Gannon and Wilson Taylor recently wrote in an April 2021 issue of British magazine *Tribune*, "[the] impressive conceit of speculative fiction, and such an experimental approach to history also offers new ways to talk about the devastating effects of trauma that Vonnegut and others experienced during that catastrophic era. It also enacts a literary strategy to ironize and reimagine ourselves, our stories, and our societies."

THE ORIGINAL SERIES

So it was too with *Star Trek*. *The Original Series* was developed just as the Vietnam War was becoming a mostly American affair. In a hint of what was to come from the franchise in the future, in dealing with contemporary social issues, *The Original Series* displaced the Vietnam War in space and time. Gene Roddenberry submitted the first outline for *Star Trek* as an action-adventure science fiction created to "keep even the most imaginative stories within the general audience's frame of reference." And, by the time the first episode was aired in September 1966, the United States was the major player in a brutal war that was savaging Indochina as America itself was being torn apart.

When the last episode of *The Original Series* was broadcast in June of 1969, the war was already looking catastrophic. In response to this growing catastrophe, a quartet of episodes in *The Original Series*, aired between the spring of 1967 and January 1969, covered the most vital period of the Vietnam War, and for America, relate specifically to the war. If we look at this quartet as a miniseries, they dramatize an arresting and disturbing evolution in the war's sway on both the franchise and the United States.

THE CITY ON THE EDGE OF FOREVER

The first of the quartet is "The City on the Edge of Forever." As we know, broadcast in the penultimate week of the franchise's first season, the episode's script went through some interesting edits regarding the question of war. Harlan Ellison's original script, written in May 1966, was simply a poignant tale of lost love. It made no reference to Kirk's love interest, the social worker and slum angel Edith Keeler, as being a peace activist, let alone refer to her peace movement that had the potential of derailing history. (Remember that Keeler dies in a traffic accident and, if she had not been killed, she would become the founder of a peace movement that will misdirect the course of history by delaying United States entry into WWII, allowing Nazi Germany time to develop atomic capability and launching V-2 rockets to conquer the globe.)

In the edited script of June 1966, Spock divines potential futures should Keeler avoid the traffic accident and live. He suggests that her pacifist philosophy was the cause of changing the war's outcome for the worst. And that's how the episode finally aired in spring of 1967. A main plot line whose subtext was the burgeoning peace movement against the war in Vietnam. Indeed, confirmation came in 1992 when Bob Justman, television producer of *Adventures of Superman*, *The Outer Limits*, and

Mission: Impossible, as well as *Star Trek*, was asked whether the episode makers purposefully recreated the script to have the contemporaneous antiwar movement as subtext. "Of course we did," Justman replied.

Prior to "The City on the Edge of Forever" being broadcast, the most incredible early domestic development of the Vietnam War was the growth of the antiwar movement. The size and passion of this peace movement were without precedent in the history of US wars. The first big antiwar demonstration took place in Washington in April 1965, just weeks after the first open posting of US combat troops to Vietnam. Soon after, a heartfelt campaign began to brew, one that aimed to educate the American people about the war's history. The campaign included the teach-in movement on college campuses, along with the publication of many books, journals, and pamphlets.

Through such publicity, millions of Americans found that, rather than the war being about the defense of a country called South Vietnam from attack by the communist country of North Vietnam, the conflict was actually a war of independence, like their own Revolutionary War against the British. Vietnam was fighting first against the French, and then against a dictatorship installed in the south in 1954 by the United States in violation of the Geneva Accords. Americans learned how several of their governments had slowly escalated a covert war into what was already becoming America's longest overseas military conflict.

Knowing this Vietnam history is vital in understanding what "The City on the Edge of Forever" was all about. As an embodiment of a dangerously misguided peace movement, Edith Keeler is to play a pivotal role in keeping the United States out of WWII long enough for the Nazis to develop a nuclear capability. Not only that, but the Nazis win the war, rule a fascist world and, heaven forbid, destroy a potentially wonderful twenty-third-century future in which the USS *Enterprise* exists! No wonder Spock tells Kirk Edith Keeler must die.

Keeler is not a diabolical character. She is simply misguided, we are led to believe. She unwittingly plays the crucial role that leads to an unwanted fascist future. Indeed, she is a visionary. Remember that she foresaw a potential future like the one portrayed by the franchise: "One day, soon, man is going to be able to harness incredible energies, maybe even the atom, energies that could ultimately hurl us to other worlds, maybe in some sort of spaceship. And the men who reach out into space will be able to find ways to feed the hungry millions of the world and to cure their diseases . . . And those are the days worth living for."

The broadcast script of "The City on the Edge of Forever" obviously became an allegory that suggested the peace movement against US involvement in Vietnam posed a danger to the course of history. Like Edith Keeler, no matter how noble their intentions, how idealistic their motivation, the peace movement was not on the right side of history. Consequently, "The City on the Edge of Forever" is *pro* Vietnam War. The story supports the "just war" theory that war, however terrible (though less terrible with the right conduct), does not have to be the worst option. Furthermore, responsibilities, undesirable outcomes, or preventable atrocities may justify war (even doing away with well-intentioned, attractive people like Keeler, who stand in the way of historical necessity), such as waging remorseless warfare against evil expansionist forces like Nazi Germany or the imagined communist empire trying to take over Indochina. (Incidentally, *Star Trek* is leaning on an ancient tradition here regarding the "just war" theory. A recent study discovered that the just war tradition can be traced back to Ancient Egypt, showing that the just war thought developed outside of Europe and existed for millennia before the advent of Christianity or even the emergence of Greco-Roman teachings.)

A PRIVATE LITTLE WAR

During 1967, the year in which "The City on the Edge of Forever" was broadcast, the peace movement against the Vietnam War was growing, and yet still represented a minority of Americans. Most expected victory in Indochina, believing that the war was not only necessary but also winnable, maybe even imminently so.

Despite the US media's near-total support of the war, Americans started to see glimpses of war's appalling reality. Though the Pulitzer Prize–winning photograph of "Napalm girl" was five years away (the famous photo shows a nine-year-old child running naked on a road after being severely burned on her back by a South Vietnamese napalm attack), napalmed children and torched villages were already being beamed into the typical American home.

In 1967, President Johnson's administration embarked upon a PR exercise on the war, massaging the truth to suggest that, after a slow escalation, the war was on the brink of success. The American people, it was suggested, needed to be patient, and the country should reject calls for troop withdrawal, as well as calls for a speedy end to the war through the use of nuclear weapons.

Into this climate came the *Star Trek* episode that deals most explicitly with the Vietnam War. "A Private Little War" was the nineteenth episode of the second season of *The Original Series*. Written by Gene Roddenberry and first broadcast on February 2, 1968, "A Private Little War" is a tale about the discovery by the crew of the Enterprise of Klingon interference in the maturing of a previously peaceful planet, an interference which develops into an arms race.

The planet in question is the pastoral idyll of Neural (just one letter away from being "Neutral"!), a world which Kirk recalls from a previous visit as so primitive and peaceful that it was almost Edenic. But war has begun on Neural, and an unequal one at that. One side of the war, those called "the villagers," have been enigmatically armed with weapons far beyond the tech level of the planet. These villagers, clearly meant to represent the US government view of the Vietnamese in the north, have been launching a kind of guerrilla warfare on the peaceful "hill people," who represent the US government view of the Vietnamese in the south. To be clear, the villagers appear to be armed with only basic handmade flintlocks. But, on further inspection, it turns out that the weapons have been mass produced by an outside imperialist power, one which has sneaked them in in a deliberate attempt to make their provenance appear indigenous.

Who would pull off such a diabolical plan? Why, the Klingons, of course. That evil empire and favorite *Star Trek* analog for China, the Soviet Union, or any other communist conspiracy. Their diabolical goal, naturally, is to conquer this primitive planet, itself an analog for Vietnam, and any part of planet Earth subject to the menace of communist expansion.

And so "A Private Little War" advanced the US government's official account of the backstory of the Vietnam War, that the conflict had started because of outside interference by an external empire of "evil," namely the Soviet Union or China. In actual fact, as many millions of Americans, including those watching *Star Trek*, were then discovering, the war had begun as a revolutionary war of independence against the existing French empire in Indochina.

Naturally, "A Private Little War" doesn't openly advocate a major escalation of the Vietnam War. Rather, it appears to warn against any form of military intervention, especially one that might lead to open warfare between America, represented in this episode by the Federation, and some evil communist empire, naturally represented by the Klingons.

The debate about the war is showcased using heated exchanges between Kirk and McCoy. It is very noteworthy that the Holmesian über-rationalist Spock, who early on in the episode declares a stark warning against messing in planet Neural's business, is then wounded and forced to recuperate on the Enterprise, conveniently omitting him from any further input into the debate and decision-making. Maybe Spock's usual role was unwelcome and potentially muddying here, given that his usual job is that of an objective analyst, an independent commentator on human affairs, which may well have ended up with the Vulcan pointing out the insane illogic of the entire business and thus denouncing the Neural/ Vietnam War.

Before his debate with McCoy, Kirk had already unilaterally decided to give military training to the hill people and to arm them with the same weapons as the villagers. McCoy is incensed by the captain's actions. And McCoy stresses the horrific potential outcome for the people of this pastoral planet, a people whom the Federation are meant to be aiding, given the Prime Directive (also known as the "non-interference directive"), which prohibits Starfleet members from interfering with the internal and natural development of alien civilizations. The Prime Directive applies especially to civilizations which are below a certain tech threshold and seeks to prevent starship crews from using their superior tech to impose their own values on them. McCoy's speech resonates strongly with Vietnam as he says, "You're condemning this whole planet to a war that may never end. It could go on for year after year, massacre after massacre."

And yet Kirk is unrepentant. He argues that he is simply setting up an equilibrium of power with a kind of "scientific" method in balancing forces. One section of the dialogue between the two makes direct reference to the Vietnam War:

McCoy: "I don't have a solution. But furnishing them with firearms is certainly not the answer!"

Kirk: "Bones, do you remember the twentieth-century brush wars on the Asian continent? Two giant powers involved, much like the Klingons and ourselves. Neither side felt that they could pull out?"

McCoy: "Yes, I remember—it went on bloody year after bloody year!"

Kirk: "But what would you have suggested? That one side arm its friends with an overpowering weapon? Mankind would never have lived to travel space if they had. No, the only solution is what happened, back then, balance of power."

McCoy: "And if the Klingons give their side even more?"

Kirk: "Then we arm our side with exactly that much more. A balance of power—the trickiest, most difficult, dirtiest game of them all—but the only one that preserves both sides!"

With his actions and words, Kirk sides here with the American government under Johnson. It's the old "end justifies the means" argument of nineteenth-century Russian revolutionary Sergey Nechayev, that if a goal is sufficiently morally important, then any method of getting it is acceptable. Kirk, like Johnson's administration, is suggesting that, though the path ahead may be protracted and perilous, a resolute determination to see things through will eventually lead out of the Vietnam quicksand and into the glorious future of starships and space travel.

The America that watched "A Private Little War" was becoming a very different country. There was growing disillusion with an apparently endless war. There was a huge polarity in political views. On the one hand there were authoritarian and militaristic voices, such as Republican senator Barry Goldwater, who supported the use of tactical nuclear weapons, and then Governor of California Ronald Reagan, who suggested Vietnam be paved over so that US troops could be brought home by Christmas. And, on the other hand, there was the colossal peace movement, which demanded the United States withdraw and let the Vietnamese people free. McCoy's objection to the war seems toothless and impotent to the point of mere moral outrage while Kirk's position, seemingly torn by his refusal to accept the two extremes outlined above, is meant to be that of a pragmatic moderate. The story of "A Private Little War" ends in a most un-*Star Trek*-like fashion. Steeped in a sense of dread and disillusion, Kirk orders Scotty to send one hundred flintlock rifles for the hill people: "a hundred serpents . . . for the garden of Eden," but when McCoy tries to placate Kirk, he says forlornly, "We're very tired, Mr. Spock. Beam us up home."

"A Private Little War" floated a hypothesis of the Vietnam War that was out of time. When Kirk maintains that keeping their presence on the planet a secret for an "enormous tactical advantage" over the Klingons, he is alluding to the period of covert US involvement of the early 1960s, and not the open warfare of 1968. Indeed, by early 1968, despite all governmental claims, the Kirk strategy of gradual escalation was on a road to nowhere and McCoy's caveat that the war may never end was not so easily refuted.

And so, the first two episodes of our quartet, "The City on the Edge of Forever" and "A Private Little War" held that the Vietnam War was a hideous necessity. Only by following the "just war" theory could mankind find itself on that yellow-brick road to the sensational twenty-third century of space travel and sex with aliens. However, a mere two days before "A Private Little War" was broadcast, the Vietnam War was about to take a very dramatic turn.

THE OMEGA GLORY

By the time "The Omega Glory" was broadcast on March 1, 1968, all expectations of victory in Vietnam had vanished, and a sense of uncertainty grew. On January 30, 1968, the Tet Offensive had begun. One of the largest military campaigns of the Vietnam War, insurgent forces of the Viet Cong and the North Vietnamese People's Army simultaneously attacked every American base and over one hundred cities and towns in South Vietnam. The crushing Tet Offensive made it crystal clear that the Vietnam War would not end in victory.

The Tet Offensive weighed heavy in its influence on "The Omega Glory." The episode was actually in the can by December 1967, by which date American antiwar media had begun to debunk governmental optimism with reports of the nation's rapidly deteriorating military situation. Written again by Gene Roddenberry, the satirically titled "The Omega Glory" had now positioned *Star Trek*'s vision closer to the antiwar movement and presented the profoundly dark consequences of the Vietnam War.

It's very much worth considering at this point one of the main themes in Francis Ford Coppola's epic Vietnam War movie, *Apocalypse Now*, widely considered one of the greatest films ever made. Detachment is one of Coppola's main themes as, during the Vietnam War, many American soldiers felt a level of detachment due to a lack of understanding as to why they were there. Some performed gross acts of violence just because they thought they were fulfilling a duty. The net outcome was that the patriotic "war hero" idea that emerged after WWII was totally annihilated for many young soldiers.

In fact, "The Omega Glory" holds a view on the war close to that expressed by McCoy in "A Private Little War," namely, it sees an endless war whose victims are no more seen as just an alien people in far-flung lands like planet Neural or Indochina. Now the imagined victim is America itself, an impoverished former civilization relegated to barbarism. The story starts when Kirk, McCoy, and Spock visit the planet Omega IV, whose appalling history becomes known to them. The world of Omega IV is ruled by a race of Asiatic villagers known as the "Kohms." The Kohms are at war with a fair-skinned race of savages known as the "Yangs," who seem so Neanderthal they are almost subhuman. And yet the Yangs are about to overthrow the Kohms by force of number and aggression. This picture is made more complex by the fact that Captain Tracey of the Federation starship USS *Exeter* has violated the Prime Directive, as Kirk did in "A Private Little War," by intervening in the world's war on the side of the Kohms, using his phasers to personally slaughter many hundreds of Yangs.

Medical research carried out by McCoy shows that there had once been an advanced civilization on Omega IV. A civilization evidently wiped out through constant warfare. After some survivors show signs that they engaged in "bacteriological warfare," McCoy comments that humans were "once foolish enough to play around with that." Ultimately, Spock concludes that Omega IV is a case of parallel evolution with Earth: "they fought the war your Earth avoided, and in this case the Asiatics won and took over the planet."

Spock's conclusion is drawn when he and Kirk finally realize the rather obvious import of the names of the opposing enemies, as Kirk declares "Yangs? Yanks. Yankees!" And Spock replies with "Kohms. Communists!" By this point in the episode, the Yangs are being goaded by Captain Tracey to execute Spock, Kirk, and McCoy. With a heavy drama, the scene sees the entrance of the sacred banner of the Yangs, a ragged old American flag, clearly the "glory" of the tale's title. Now ignorant of the truth and justice for which they were meant to be fighting in their war against the Communists, these Yankees have become feral savages, reeling toward the insane and bestial.

The question of insanity is another main theme in Coppola's *Apocalypse Now*. The question underpins Willard's entire mission against Colonel Kurtz. Willard is told early in the movie regarding Kurtz's crimes that "Every man has got a breaking point. You and I have. Walter Kurtz has reached his. And very obviously, he has gone insane," and that Kurtz's methods had become "unsound." In "The Omega Glory," all that remains of the Yankees' original ideals are their words of worship, rattled-off versions of the Pledge of Allegiance, and the preamble of the US Constitution, mouthed like a meaningless mantra.

In a typically stagy episode's end, Kirk snatches the Yangs' cherished copy of the preamble to the US Constitution. He then recites it, placing extra weight on "We the People." Kirk then lectures the Yangs, who now revere him as a god due to the "miraculous" materialization of an Enterprise rescue team, that the preamble was "not written only for the Yangs, but for the Kohms as well." Realizing that such a position must seem like outrageous dissidence to the Yangs, Kirk soldiers on: "They must apply to everyone, or they mean nothing." The countenance of the Yangs slowly becomes more humane and Kirk appears to derive satisfaction that he has stirred the Yangs from an age of mindless anti-Communist warfare. Perhaps now, to the sight of Old Glory and the sound of the "Star-Spangled Banner," this planet may also return to a more idealistic road. If only real life were so simple.

Thus, the third episode in our quartet, "The Omega Glory," held that this war in Indochina would never be winnable, had little prospect of an ending, and may even develop into a seemingly endless and mutually ruinous conflict between the "Yanks" and the "Commies," one which had the potential of wiping out civilization and humanity itself. The ideal of what it is to be a true patriot is depicted as being the opposite to anti-Communism and militarism, even if it does contradictorily use jingoistic totems such as Old Glory and the "Star-Spangled Banner" to make its point about the brotherhood of man.

Kirk even suggests that the journey be taken further, with his focus on "We the People." He seems to suggest that Americans should make their collective voice heard more clearly in the nation's affairs. In other words, the journey of the war has provided lessons to be learned at home, as well as abroad.

This question of a journey is another theme in Coppola's *Apocalypse Now*. Coppola structured the story of his famous movie as a journey. The main protagonist arrives in Indochina with the aim of being given a new mission because after his first tour of Vietnam he no longer fits into his civilian life. And so, his journey to Kurtz is literal and metaphorical—a journey into the jungle and into himself.

LET THAT BE YOUR LAST BATTLEFIELD

Like the nation itself, science fiction writers were still sharply divided about the Vietnam War, but *Star Trek* now stood on an antiwar platform. In the pages of America's leading sci-fi magazines in early 1968, more than one hundred and fifty writers bought advertising space either backing or condemning the continuing conflict. The advertisements were first published in the March issue of the *Magazine of Fantasy and Science Fiction*. Not a single soul associated with *Star Trek* was among the seventy-two who signed the ad that said "We the undersigned believe the US must remain in Vietnam to fulfill its responsibilities to the people of that country." On the contrary, *Star Trek* writers Harlan Ellison, Jerry Sohl, Jerome Bixby, and even Gene Roddenberry himself were among the eighty-two who signed the opposing ad: "We oppose the participation of the US in the war in Vietnam."

Indeed, 1968 was a year of global protest, not just for science fiction writers. In reaction to the Tet Offensive, protests also sparked a broad movement in opposition to the Vietnam War all over the United States as well as in London, Paris, Berlin, and Rome. There was a worldwide upsurge in social conflicts, typically popular rebellions against the military and the bureaucracy. In America, these protests peaked after the assassination of Dr. Martin Luther King, Jr. Huge uprisings erupted in one hundred and twenty-five cities in a single week, and more than 55,000 troops had to augment police to quell the uprisings. Even the US Capitol had to be defended by combat troops, while way above Washington columns of black smoke rose from burning buildings. These protests marked a turning point for the civil rights movement, which produced revolutionary movements like the Black Panther Party. Richard Nixon won the 1968 presidential election as a peace candidate. During his campaign, Nixon said "as we look at America, we see cities enveloped in smoke and flame . . . I pledge to you tonight that the first priority foreign policy objective of our next administration will be to bring an honorable end to the war in Vietnam."

Against this global backdrop, in January 1969, just days before Nixon's inauguration, and still four years before the end of official American involvement in the Vietnam War, *Star Trek* aired the appropriately named episode "Let That Be Your Last Battlefield." The story that unfolds is a parable of two races on an alien world. Each race is curiously black on one side of the body and white on the other, and they wipe out one another in an escalating violent conflict. The "master" race, white on the left, black on the right, enslave and exploit the other race, who differ only in the sense that they're black on the left and white on the right.

"Let That Be Your Last Battlefield" casts a satirical light on America's racial conflict of the 1960s. In a reference to the inordinate number of deaths being suffered by black Americans in Vietnam, the leader of the oppressed alien race asks crew members of the Enterprise, "Do you know what it would be like to be dragged out of your hovel into a war on another planet, a battle that will serve your oppressor and bring death to your brothers?"

The final endpoint of such mutual racial hatred becomes clear. When the Enterprise reaches their home world, Spock detects that there are no longer any sapient life forms present on their planet: "they have annihilated each other totally." As a sole representative of each warring race still fights to the end, a backdrop reveals actual footage of burning US cities. The prophecy of global disaster from the Vietnam conflict suggested in "The Omega Glory" has come home to roost.

And so our quartet comes full circle. In its response to the political science of such a monstrous war, *Star Trek*'s first two episodes, "The City on the Edge of Forever" and "A Private Little War," held that the Vietnam War was simply a necessary means to the end of a glorious future dramatized by *Star Trek*. And yet the last two of the quartet, "The Omega Glory" and "Let That Be Your Last Battlefield," are so steeped in the radicalism of the times that they call for a revolutionary change of course—an end to wars abroad, and an end to the racial war at home. Only this new course would bring us the multicultural future of the USS *Enterprise*.

WHY IS "DARMOK AND JALAD AT TANAGRA" SO IMPORTANT TO SCIENCE?

Picard: "The Enterprise is en route to the uninhabited El-Adrel system. Its location is near the territory occupied by an enigmatic race known as 'The Children of Tama' . . . are they truly incomprehensible? In my experience, communication is a matter of patience, imagination. I would like to believe that these are qualities that we have in sufficient measure."
—Joe Menosky, teleplay of *Star Trek: The Next Generation*,
"Darmok" (1991)

"Occasionally, I get a letter from someone who is in 'contact' with extraterrestrials. I am invited to 'ask them anything.' And so over the years I've prepared a little list of questions. The extraterrestrials are very advanced, remember. So I ask things like, 'Please provide a short proof of Fermat's Last Theorem.' Or the Goldbach Conjecture. And then I have to explain what these are, because extraterrestrials will not call it Fermat's Last Theorem. So I write out the simple equation with the exponents. I never get an answer. On the other hand, if I ask something like 'Should we be good?' I almost always get an answer."
—Carl Sagan, *The Demon-Haunted World* (1995)

DARMOK

In "Darmok," the 102nd episode of *Star Trek: The Next Generation*, Enterprise assumes orbit around the planet El-Adrel IV, where a Tamarian vessel has been broadcasting a mathematical signal for weeks. The Federation had been in touch with the "Children of Tama" previously, but neither side could make themselves understood, despite the fact that a universal translator interpreted the actual Tamarian words. The Tamarians, an apparently peaceable and technologically advanced race, communicate through the use of metaphoric allusions to their history and mythology, so a cultural knowledge of the Children of Tama is needed to understand their thoughts and intentions. Similarly, the Tamarians find great difficulty in understanding Picard's far more direct use of language.

On hailing the alien ship upon arrival, contact with the Tamarian vessel proves more difficult than the optimistic Picard first predicted. Indeed, before we proceed to talk about the scientific importance of this story, it would be wise to look at the way in which the Children of Tama actually communicate. Dathon, the Tamarian captain, begins with "Rai and Jiri at Lungha. Rai of Lowani. Lowani under two Moons. Jiri of Umbaya. Umbaya of crossed roads. At Lungha. Lungha, her sky gray." After translation, Data offers the explanation, "The Tamarian seems to be stating the proper names of individuals and locations," to which Picard replies, "Yes, but what does it all mean?"

After Picard's staid reply causes confusion among the Tamarians, their vessel unexpectedly transports its captain and Picard down to the surface of El-Adrel IV below. Captain Dathon has with him two Tamarian daggers which carry some ritual gravitas. The crew of Enterprise try to bring Picard back on board, but the Tamarians are one step ahead: they induced a particle scattering field in the ionosphere of El-Adrel IV to make teleportation no longer possible.

MEANWHILE, DOWN ON EL-ADREL IV

Down on the world below, Dathon throws one of the daggers to Picard who, perhaps understandably, thinks he's being provoked into a fight. Up on the Enterprise, Riker comes to the same conclusion and tries to contact his Tamarian counterpart, only to be seemingly stonewalled with

"Darmok at Tanagra." Riker returns with the aggressive reply of "Your action could be interpreted as an act of war!" To which his counterpart complains to his Tamarian crew, "Kiteo, his eyes closed," before replying to Riker "Chenza, at court. The court of silence," and closes the channel.

Darkness draws in on this alien world, and as Picard has trouble making sparks catch a flame, Dathon basks in his burning blaze of a fire. So Dathon tosses Picard a torch, saying "Temba." Picard at first thinks Temba means fire, until Dathon says "Temba, his arms wide." Then the penny starts to drop with Picard. He says "Temba is a person. His arms wide . . . because he's . . . he's holding them apart. In, in . . . generosity. In giving. In taking. Thank you."

Dawn breaks, and Picard is awoken with some urgency by Dathon who says "Darmok! Darmok and Jalad at Tanagra." But Picard is still stumped. Suddenly, a portentous and monstrous roar is heard in the distance. Picard now gladly takes the weapon offered by Dathon. Picard makes to run, but Dathon assures him "Shaka, when the walls fell." Picard has another epiphany. "Shaka. You said that before. When I was trying to build a fire. Is that a failure? An inability to do something?"

The monster nears, and Dathon tries to take control of the imminent battle. "Uzani, his army at Lashmir," he says, to which Picard replies "At Lashmir? Was it like this at Lashmir? A similar situation to the one we're facing here?" Then Dathon says "Uzani, his army with fists open," and Picard responds with "A strategy? With fists open?" When Dathon then says "His army, with fists closed," Picard realizes the meaning and, showing that he understands, says "With fists closed. An army, with fists open, to lure the enemy . . . with fists closed, to attack? That's how you communicate, isn't it? By citing example, by metaphor! Uzani's army, with fists open." Dathon, realizing that Picard's finally got the gist replies with "Sokath! His eyes uncovered!"

Armed with this plan, the pair proceed to execute it, but just as Picard is on the verge of confusing the monster so that Dathon can attack, Enterprise tries to retrieve Picard, so the plan fails and the monster mauls Dathon. Picard rematerializes, running to Dathon who struggles to say "Shaka," as this time Picard finishes the metaphor "when the walls fell."

TAMARIAN COMMUNICATIONS

Meanwhile, Enterprise-based Troi and Data gain some ground deciphering Tamarian communications. Troi points out that the Tamarian ego structure doesn't seem to allow for self-identity, at least not in usual parameters. Indeed, their ability to abstract is so unusual that they communicate through narrative imagery. In short, they speak by reference to the people and places that appear in their mytho-historical accounts. As their analysis continues, Troi, Data, and also Crusher conclude that even given their new understanding, without knowing the mythical origins of the people that feature in the Tamarian language, they have scant hope of grokking the speech of their alien friends.

Down on El-Adrel IV, Picard begins to better understand the intention of the Tamarian captain in taking them both down to the planet's surface: "You hoped that something like this would happen, didn't you? You knew there was a dangerous creature on this planet and you knew, from the Tale of Darmok, that a danger shared, might sometimes bring two people together. Darmok and Jalad at Tanagra. You and me, here, at El-Adrel."

Dathon slowly suffers for his meeting with the monster, so Picard adopts the method of Tamarian syntax by telling the Earthly tale of Gilgamesh: "Gilgamesh, a king. Gilgamesh, a king. At Uruk. He tormented his subjects. He made them angry. They cried out aloud, 'Send us a companion for our king. Spare us from his madness.' Enkidu, a wild man . . . from the forest, entered the city. They fought in the temple. They fought in the streets. Gilgamesh defeated Enkidu. They became great friends. Gilgamesh and Enkidu at Uruk." Finally, Dathon succumbs to his injuries and ultimately Picard is back on the bridge of the Enterprise complete with a new linguistic understanding.

THE ANTHROPOLOGY OF INTERSTELLAR COMMUNICATION

Why is the "Darmok" episode of *Star Trek: The Next Generation* so important to science? Well, ever since April 8, 1960, when American astronomer Frank Drake inaugurated a new era in the search for alien civilizations beyond Earth, scientists have not only been searching for a broadcast frequency that they hoped would be a universal meeting place with aliens, they've also been searching for a universal language, one which might mean that humans are on the same "wavelength" as a listening extraterrestrial civilization. And that's where anthropology comes in.

Anthropology is the study of the human race, its culture and society, so it would be a fair question to ask why on Earth anthropology should apply to the alien life debate, which clearly deals with nonhumans. When the modern era of the Search for Extraterrestrial Intelligence (SETI) began, the discipline of anthropology was about a century old. SETI scientists realized that the methods anthropology had developed for analyzing cultures and cultural evolution could also be applied to any intelligent species that may exist beyond the Earth. If such an alien culture was millions of years old, its associated cultural evolution would have implications for beneficial mutual interactions between humans and aliens.

As the SETI program aims to search for intelligent alien life, especially through monitoring signs of transmissions from civilizations on other planets, the cultural and anthropological aspects of that communication would be of equal importance. Throughout SETI's history, social scientists commonly contributed to scientific conferences where alien intelligence and communication were discussed. For example, when NASA sponsored a 1972 symposium at Boston University titled "Life Beyond Earth and the Mind of Man," anthropologist Ashley Montagu argued, as if predicting some of the themes in "Darmok," that "it is the communication we make at our initial encounter that is crucial." Montagu continued, "I do not think we should wait until the encounter occurs; we should do all in our power to prepare ourselves for it. The manner in which we first meet may determine the character of all our subsequent relations. Let us never forget the fatal impact we have had upon innumerable peoples on this Earth—peoples of our own species who trusted us, befriended us, and whom we destroyed by our thoughtlessness and insensitivity to their needs and vulnerabilities."

You can understand why Montagu argued for the study of intercultural contact. A brief review of the history of some of the more notorious instances of European and American colonialism richly informs any discussion of first contact between humans and "aliens." For centuries, European societies had few tech advantages over the developed societies of the East, especially China, as Asian cultures had superior tech until relatively recently. But, as the early modern period began, Europeans started to push forward in areas such as shipbuilding, map-making and navigation, and the production of ordnance (remember that great ocean-going ships were built in China decades earlier, and that tech invention such as gunpowder, the compass, and the stern-post rudder actually came from China originally). By marrying up naval and military progress with state sponsorship and ruthless intentions, the West forced its will upon "alien" cultures and appropriated a huge amount of wealth. One of the many things missing from these historical cultural clashes was a reciprocity of understanding in human interaction.

RECIPROCITY OF COMMUNICATION

"Darmok" is a good example of such reciprocity. In fact, reciprocity is forced upon Picard, who is abducted and marooned on the planet so that he simply must try to communicate. Reciprocity is a basic principle of human interaction. And "Darmok" is saying it would also be so for any sentient race, such as the Tamarians. What kind of encounter or contact can we humans expect with aliens in the future?

Well, the famous "close encounters" classification of alien contact was developed by American astronomy professor Josef Allen Hynek. A close encounter of the first kind was a visual sighting of an unidentified flying object, apparently less than 500 feet away. A close encounter of the second kind was a physical effect, like an electronic interference, a bodily effect like paralysis, or a physical trace such as scorched vegetation. And a close encounter of the third kind was the presence of an alien creature, such as a humanoid, or even a robot. But if contact is to happen at all, between Earth and extraterrestrial intelligence, it's more likely to occur across vast interstellar distances, on time scales of decades, centuries, or millennia.

COMMUNICATION WITH EXTRATERRESTRIAL INTELLIGENCE (CETI)

CETI, the Communication with ExtraTerrestrial Intelligence, is a branch of SETI. CETI focuses on composing and deciphering interstellar messages that theoretically could be understood by another technological civilization. Partly using anthropology to assess the rationale and procedures for attempting to establish contact, this ambitious mission to communicate with the ultimate "Other" is as exotic and thought-provoking an exercise as "Darmok" suggests. For instance, according to many CETI experts, contactable alien civilizations would be many light-years away, and light-years ahead of us, so that the prospect of meaningful conversations with such advanced and distant alien civilizations would be rather limited.

Solutions to this limitation revolve around two assumptions. First, that advanced alien civilizations capable of radio communication would share with humans the same fundamental methodologies, use similar physics at least as well as us. Second, that those aliens wishing to make radio contact would avoid "natural" languages and create artificial ones, presumably based on common methodologies and scientific knowledge.

TALKING TO ET

The research of CETI has centered around four main areas: math languages, picture messages (like the famous Arecibo message of basic data about Earth and humanity sent to star cluster M13 in 1974), algorithmic communications, and computational methods for detecting and decoding so-called natural language communication. In human history, between civilizations old and new, many writing systems remain undeciphered. For example, Linear A is a writing system used by the Minoan civilization on Crete between 1800 and 1450 BCE. It is hypothesized that Linear A was used to write the Minoan language, and was the main script used in palace and religious writings of the Minoan culture. (Incidentally, the word "linear" is a reference to the fact that the script was inscribed using a stylus, which cut lines into clay tablets, in contrast to cuneiform, which was written by pressing wedges into the clay.) As CETI experts can attest, the only part of Linear A so far deciphered with any confidence is the signs for numbers, and yet even those are known only in their numerical form—the actual words for those same numbers remain a mystery.

Much of the CETI research effort is focused on how we surmount similar issues with deciphering messages that may arise in interplanetary communication. In short, scientists need to work out not just what we'd say when ET calls, but also how we would go about saying it. Such communication had a less than auspicious start. From the nineteenth century comes the mildly apocryphal tale that Austrian astronomer Joseph Johann von Littrow came up with the bright idea of digging huge trenches in the Sahara desert, filling them up with water, then pouring kerosene on top. The resulting conflagration, it was hoped, would signal to any neighboring ET that a very adept and sophisticated human civilization was sending the flaming message.

LANGUAGE OF THE COSMOS

The flaming Sahara scheme never materialized. Not only is water rather difficult to come by in the Sahara, but there was, and continues to be, a distinct lack of Martians to notice the fiery signal. But come the 1960s, a truly sophisticated idea came along in Lincos. Dr. Hans Freudenthal's book *Lincos: Design of a Language for Cosmic Intercourse* founded the field of exo-linguistics, becoming the first artificial language made for the purpose of communicating with alien life.

Lincos, an abbreviation of the Latin phrase *lingua cosmica*, is nonetheless a language that uses math to transmit the minutiae of daily terrestrial life as well as universal truths. Its text is hardly user-friendly, being made up of dense, technical jargon and math formulas that are all mostly incomprehensible to the ordinary reader. The aim of Lincos is to provide a language that can be understood by a technical "person" from the get-go. It doesn't need acquaintance with natural languages, not even their syntactic structures.

Lincos is a spoken language rather than a written one. It's comprised not of letters but of phonemes, units of sound that differentiate one word from another, and is governed by phonetics rather than spelling. The speech is comprised of unmodulated radio waves, which vary in duration and length, encoded with a mishmash of symbols taken from science, math, logic, and Latin. When combined, these waves can communicate a range of knowledge, from fundamental math formulae to philosophical ideas of love and death.

LEARNING LINCOS

How does Lincos work in practice? The first part of a sent message should include numerals that introduce the receiver to the message's version of math. This would consist of brief, regular pulses, with the number of pulses correlating to the specific numeral; one pulse for 1, two pulses for 2, etcetera. The second part of the message would beam out basic formulae, including the use of symbols such as =, +, or > to show the properties of notation and math relations (for example: ">" to confirm that seven is more than six). Successive parts of the message would gradually build up in complexity, progressing from numerals to formulae to more sophisticated subjects like human culture.

Of course, as "Darmok" testifies, despite the mathematical and logical nature of Lincos, extraterrestrials may not think like us at all, in which case the Lincos logic would be lost on them, as Picard's initial overtures were similarly lost on the Tamarians. And so even something as well thought out as Lincos requires that the message receiver is to some degree humanlike in their mental state. And it might well prove impossible to communicate with extraterrestrials who do not fulfill this requirement.

Nonetheless, some SETI scientists believe that, if there are intelligent aliens in the cosmos, the chances are that they will think like us, or will be familiar with a form of math like ours. The argument goes something like this: if an extraterrestrial civilization can build a receiver for interstellar messages like ours, then they know the science and math needed to make such a machine. The famous American cognitive scientist Marvin Minsky put it another way. Minsky suggested that like-minded intelligent aliens were subject to the same ultimate constraints—limitations on space, time, and materials.

LINCOS EVPATORIA MESSAGES

In 1999, astrophysicists sent out into the cosmos a series of Lincos messages from a radio telescope in Ukraine. The transmissions, known as the Evpatoria Messages, were the third to ever be sent out to potential extraterrestrial civilizations. Previous messages sent to the stars had been pictorial, such as the Arecibo message in 1974. But the Evpatoria Messages were encyclopedic, containing as much information about life on Earth as possible.

The astrophysicists created a character alphabet that would enable a maximum amount of data to be transmitted during their limited use of the telescope. The messages began by teaching the recipient how to count, then continued with data that should be known to any sentience in the cosmos, progressing to more complex topics such as physics and biology. Here is an image of the first page of the Evpatoria Messages. It's taken from a September 26, 2016, article in *Smithsonian* magazine by Yvan Dutil and

First page of the Evpatoria Message (Yvan Dutil and Stéphane Dumas)

Stéphane Dumas. They explain how this first page "defines the numbers used in the message. It lists the numbers 0 to 20, reading across and omitting several of them. Each number is given in three forms: as a group of dots, as a binary number, and as a symbol in base-10 format."

The overall message was twenty-three pages long, with each page 127 × 127 pixels, and was beamed to its targets three times all told, in order to ensure sufficient redundancy. The message also contained a request for a reply from its alien recipient. The target star systems for the Evpatoria Messages were chosen from SETI's list of potential extrasolar systems that might harbor sentient life. These candidate systems, located between fifty and seventy light-years from Earth, were selected according to a number of criteria, including the age of the star and its position in our Galaxy.

But don't hold your breath for a Darmok-like exchange with an alien race any time soon. The first Evpatoria message will reach its cosmic destination (address at Hip4872 in the constellation Cassiopeia) around the year 2035 and, if a prompt reply is forthcoming, we can expect to hear from them around the year 2058, by which time I shall be one hundred years old. Or, more likely, dead.

A TAMARIAN TWIST

Since 1999, SETI scientists have improved upon Lincos, seeking ways to refine the potential for extraterrestrial communication. One example is CosmicOS, a language created by MIT. CosmicOS is a computer program which can be triggered by the aliens who receive the message. There's also second-generation lingua cosmica, which uses constructive logic to compose the message. Of these two refinements, CosmicOS is able to beam out more data, but thus also has more logistical issues, while the new Lincos has a smaller and more workable data package, but also a more limited amount of data that can be sent. And yet these two new approaches to cosmic communications could be parts of a greater whole.

Whatever the method of approach, the future of any communication with extraterrestrial civilizations will be based solidly in math. Math is the best metalanguage to build a bridge between the artificial language used in the messages and the alien's own language. Italian physicist Galileo Galilei is often attributed with the quote "Mathematics is the language in which God has written the Universe." Given that math is the basis of science, any alien civilization that has built a listening device knows science and knows math. Indeed, like humans and Tamarians in "Darmok," the Galaxy's sentient species may have little in common but math.

All of which brings us back to the very start of that 102nd episode of *Star Trek: The Next Generation* known as "Darmok." As we can see from the script, Enterprise is responding to a signal that the Tamarian vessel has been broadcasting for weeks. As Picard says, "Apparently the Tamarians arrived at El-Adrel IV nearly three weeks ago. They have been transmitting a subspace signal toward Federation space ever since." Upon which Data replies, "The signal is a standard mathematical progression. It does not carry a specific message," and Riker concludes, "But they wanted us to know they were there." Given they know math, it all makes one wonder if Dathon's life might have been spared if the Tamarians had simply used their own version of Lincos in the first place!

ARE WE BORG?

"Cyborg is a word made up from cybernetics and organism. Organism is a word for any living creature. And cybernetics is the study of how humans (or aliens and machines) communicate and control information."
—Mark Brake, *The Science of Science Fiction* (2018)

"Power is in tearing human minds to pieces and putting them together again in new shapes of your own choosing."
—George Orwell, *Nineteen Eighty-Four* (1948)

CYBORGS

Cyborgs are a science fiction staple. You only have to think of cinema's most famous villain in Darth Vader, or the demonic Daleks in *Doctor Who*, to know that cyborgs are often sci-fi's most famous villains. Indeed, apart from the Borg in *Star Trek*, *Doctor Who* has arguably produced one of the most prominent of cyborg incarnations in the form of the Cybermen.

This cyborg business goes a long way back. One of the more unlikely early examples of a cyborg in fiction is the Tin Man from Frank L. Baum's 1900 classic book *The Wizard of Oz*, along with the movie versions that followed. The Tin Man was Dorothy's hero and companion on the yellow-brick road. What makes the Tin Man a cyborg? In the original tale, he was a lumberjack called Nick Chopper (seriously). He was engaged to a munchkin girl named Nimmie Amee (you just can't make this stuff up; actually, you can make it up, as Baum clearly did). Anyhow, the Wicked Witch of the East conjures up a magic axe that chops off his limbs one by one (these old fairy tales are often as dark as the Borg). He gradually

gets them replaced with metal versions and becomes (ta-da!) a type of early cyborg.

So, it seems, Cyborgs abide. The book that made cyborgs famous in the United States was a novel called, unsurprisingly, *Cyborg*. Written in 1972 by Martin Caidan, most folks will know *Cyborg* for the name of its hero, Steve Austin. Austin was a fictional test pilot who had a near-fatal air crash and had large parts of his body replaced with bionic limbs. For most of the mid-1970s, Steve Austin was famous on American television as *The Six Million Dollar Man*, a reference to how much it cost to rebuild him into a cyborg.

THE BORG

Star Trek, as you can see, had a long science fiction history to call upon when it invented the Borg. Let's take the example of the Cybermen from *Doctor Who*. Appearing on British television way back in 1966, the Cybermen were a fictional race of cyborgs, a totally organic species to start with, who began to implant more and more artificial parts to help them survive. They are said to originate from Earth's twin planet Mondas (this is totally made up, as we don't actually have a twin planet, last time we looked). As the Cybermen added more and more cyber parts, they became more coldly logical, calculating, and less human. As every emotion is deleted from their minds, they become less man, and more machine. Quite ingenious, really, as you can also use the idea of Cybermen to portray what we humans might one day become, if we base all our decisions on cold calculation and ignore our more human and emotional aspects. In short, we become Borg.

Now, those of you intimately familiar with the Borg will already see some similarities between the Borg and the Cybermen. In *Star Trek*, the Borg are a group linked in a hive mind known as "the Collective." They appropriate the science and tech of other species to the Collective through a process they call "assimilation." Like the Cybermen, the Borg assimilate others by a forced transformation of an individual being into a drone-like state with a surgical augmentation of cybernetic parts. And, like the Cybermen, the Borg are a fitting antagonist with their scary appearance, their huge power, and their sinister motives.

THE BORG AS BIG BROTHER

The Borg's goal is meant to be "achieving perfection." But perhaps a more interesting question is what their depiction in *The Next Generation* is meant to convey to the audience. They have become a symbol for any juggernaut against which "resistance is futile," a common phrase uttered by the Borg themselves. So, what does the Borg, this collectivist enemy par excellence, truly represent? Interpretations have, of course, included that old chestnut, "communism," while cases could also be made for religious fanaticism, the basic fear of being consumed, and globalization.

There is another science fiction influence behind the portrayal of the Borg. The machinery of Borg oppression might come from their tech, but the philosophy of oppression owes much to George Orwell's 1948 novel *Nineteen Eighty-Four*. Few sci-fi novels have been as prophetic as Orwell's haunting specter of totalitarian power. No novel written in the twentieth century has captured the popular imagination like *Nineteen Eighty-Four*. Big Brother is as famous as Frankenstein. Newspeak is as forked tongue as fake news. The legions of fictional Thought Police who read your mind are early echoes of both the Borg and of the twenty-first-century realm of fake news, political spin, and euphemism. War is conflict. Civilian casualties are described as "collateral damage." Lying politicians are now "misspeaking," or being "economical with the truth," or simply and quite clearly don't care. The very title of Orwell's classic dystopia became a cultural catchphrase. And the word "Orwellian" still ominously speaks of matters hostile to a free society. The Borg are Orwellian.

RESISTANCE IS FUTILE

The key to Orwell's story was his belief in a "catastrophic" future. It is a future of boundless despair. The book confronts the prospect of a totalitarian future that brings history to a standstill. Big Brother is unassailable: "If you want a picture of the future, imagine a boot stamping on a human face—forever." Resistance is futile. As with the Borg, Orwell's book became a mirror of the fears and frustrations of the individual caught up in a complex, overly rationalized society. It was a prophecy of a totalitarian future based on the implied aims of industrial civilization. Orwell was troubled by the devastating effects of science and technology: "Barring wars and unforeseen disasters, the future is envisaged as an ever more rapid march of mechanical progress; machines to save work, machines to save thought, machines to save pain, hygiene, efficiency, organization . . . until finally you land up in the by now familiar Wellsian Utopia, aptly caricatured by Huxley in *Brave New World*, the paradise of little fat men."

In his 1932 novel *Brave New World*, Aldous Huxley imagined a technology that titillated. Orwell foresaw the technology of control. And that's what the Borg are all about, the technology of control. The insidious nature of *Nineteen Eighty-Four*'s culture of surveillance stems from its telescreens and Thought Police. The insidious nature of the Borg's surveillance is assimilation into the collective of the hive mind. In one of Orwell's most marvelous quotes, he summed up the technology of control in the following way: "The Beehive State is upon us, the individual will be stamped out of existence; the future is with the holiday camp, the doodlebug and the secret police." The Borg are the Beehive State.

In *Nineteen Eighty-Four*, science's mastery of the machine is so complete that utopia is possible. But poverty and inequality are maintained as a means of sadistic control. The visual medium of monitoring in the two-way telescreen is a brilliant evocation of the all-seeing eye. The Borg don't need tele-screens. The Borg are the Thought Police. In Orwell's book, society is a technological nightmare. As Winston Smith dutifully follows the daily exercises on the telescreen, he is at the same time observed by it. When someone is assimilated into the Borg collective, they are at the same time observed and totally controlled by it.

ARE WE BORG?

Another fascinating question is whether humanity is becoming more Borg-like. Now, we don't mean with regard to whether we may have any cyborgs in our families, such as grandpa in glasses (though admittedly he may well be a mean fusion of man and machine). We're talking here more about the hive mind, and the kind of totalitarian control Orwell was speaking about in *Nineteen Eighty-Four*. Remember that Orwell's Party of Big Brother faked news, rationalized language, and perverted history. Time was tampered with, dates of events forgotten or unascertainable. The tech of information was used to maintain political control, underlining Orwell's point that "The really frightening thing about totalitarianism is not that it commits 'atrocities' but that it attacks the concept of objective truth: it claims to control the past as well as the future." Orwell had identified a new dark age. He saw the necessity for a social side to technological progress and to "reinstate the belief in human brotherhood."

The apparent lack of human brotherhood is an increasingly startling aspect of modern human life. On UK's Channel 4 News on June 23, 2021, Yale history professor Timothy Snyder was asked whether politicians were using propaganda techniques from Nazi Germany. Techniques guaranteed to work against human brotherhood. (Orwell was critical of the idea that a centrally planned meritocracy which would advance science was necessarily a good thing, arguing "Modern [Nazi] Germany is far more scientific than England, and far more barbarous.") Professor Snyder's reply was fascinating: "The reason why we know fascism is possible is because it already happened once, and it happened in places that are not so distant from us, either in place or in time. It's patently clear that some of the people who're involved in current politics are people who have learned from the 1920s and 1930s and are borrowing some of the tactics of the 1920s and 1930s."

The basic tactic which modern politicians are borrowing, says Professor Snyder, is a rhetorical one. He bases this idea on a notorious manual to propaganda which was composed in a Munich prison at the beginning of 1924. The manual advises that what you should do in political propaganda is always find simple slogans and repeat them over and over again with the effect of dividing your listeners into us and them. Professor Snyder says that this has clearly been revived as a tactic on both sides of the Atlantic. ("Make America Great Again," "Get Brexit Done," "Build the Wall and Crime Will Fall," that kind of thing.)

Slogans akin to "resistance is futile" are gaining traction. Modern discourse in political science is less about reasoned debate toward constructive policy, and more about friends and enemies. You're either in the Borg or not! This friends and enemies opposition is a basic fascist idea articulated by maybe the most famous Nazi theorist, Carl Schmitt. When Professor Snyder was asked what's wrong with what modern politicians are doing, given that they reflect the popular and democratic will, his reply was fundamentally about human law and existence: "It's the essence of the tradition of Anglo-Saxon law, going back to the Magna Carta. There's a reason why we have law and the reason why we have law is that law comes before the king, or in modern times law comes before the ruler, which means law comes before whatever momentary urges the ruler might say that he is embodying." So, Professor Snyder suggests, if we say whatever we think at this particular moment is what goes, that means we're saying goodbye to law, goodbye to predictability, goodbye to the basis of the system that we have had. Goodbye to human history.

HYPERNORMALIZATION

The Party of Big Brother also perverted history. In a 2016 BBC documentary called *Hypernormalization*, British filmmaker Adam Curtis argues that governments, financiers, and technological utopians have, since the 1970s, given up on the complicated "real world" and built a simpler "fake world" run by tech corporations and kept stable by politicians. *Hypernormalization* begins with this intriguing introduction: "This film will tell the story of how we got to this strange place. It is about how, over the past forty years, politicians, financiers, and technological utopians, rather than face up to the real complexities of the world, retreated. Instead, they constructed a simpler version of the world in order to hang on to power. And as this fake world grew, all of us went along with it, because the simplicity was reassuring. Even those who thought they were attacking the system—the radicals, the artists, the musicians, and our whole counterculture—actually became part of the trickery, because they, too, had retreated into the make-believe world, which is why their opposition has no effect and nothing ever changes."

Indeed, evidence suggests that the world is ripe for change. The new academic field of Cliodynamics claims to enable its scholars to analyze history in the hopes of finding patterns they can then use to map out the future. Experts in the field have found a pattern of social unrest. Their technique uses data from many civilizations, including dynastic China, ancient Rome, medieval England, France, Russia, and the United States. Analysis clearly shows one-hundred-year waves of instability. And, superimposed on each wave, there's an additional fifty-year cycle of widespread political violence. China seems to escape the fifty-year cycles of violence. But the United States does not. The violent cycles have social inequality at their root. Discontent builds up over a period of time and then that pressure is violently released. Scholars have mapped the way that social inequality creeps up over the decades, so much so that a breaking point is reached. A little late, reforms are finally made. But, over time, those reforms are reversed, and society lurches back to a state of heightening social inequality. Sound familiar? The severity of the violent spikes depends on how governments cater to the crisis. For example, the United States was in a prerevolutionary crisis in the 1910s. But a sheer drop in violence followed due to a more progressive political era. The ruling classes made calls to rein in corporations and allowed workers vital reforms. Such policies reduced the pressure and prevented revolution. Likewise, nineteenth-century Britain was able to avoid the kind of violent revolution that happened in France by making meager amelioratory reforms. However, the usual way for the cycle to resolve itself is through violence. Cliodynamics experts predict a wave of widespread violence in the 2020s, including riots and revolution.

So, change is overdue. And yet, as a result of a "fake world" run by tech corporations, Adam Curtis claims that contemporary humans lack a picture of the future. We have a system we are unhappy with. We know it's corrupt, we know that cracks are starting to show, and that the world is very often quite fake, particularly with our politicians. And yet most of us have no other coherent picture of the future. Furthermore, the manufactured system of the Internet doesn't help. It's wonderful in other ways and handy as an organizing tool, but not the best platform for presenting a coherent way out of this mess, a pathway to a brighter future.

POLITICAL TECHNOLOGISTS

All this suggests the semi-Borg-like state of contemporary consciousness, technology used to induce a kind of hive mind and control. How is human compliance achieved? Adam Curtis explains the backstory of compliance in *Hypernormalization*. In Russia there emerged a group of men who had seen how the contemporary lack of belief in politics, hardly surprising after decades of Stalinism, along with a dark uncertainty about the future, could be used to their advantage. And so they began by engineering politics into a theater of the absurd, where no one knew the difference between what was true and what was fake.

These political technologists were the key figures behind Russia's president. They had kept Putin in power, unchallenged, for fifteen years. Some of the technologists were dissidents in the 1970s and had been hugely influenced by the sci-fi writings of the Russian authors known as the Strugatsky brothers. (We previously mentioned the Strugatsky's novel *Roadside Picnic*, translated into English in 1977, when we were talking about the *Star Trek* episode "The Chase.") Twenty years later, when Russia disintegrated during the end days of Soviet communism, the political technologists got organized and took control of the media. From their positions in the media, they were able to manipulate the electorate on a huge scale. As in Orwell's *Nineteen Eighty-Four*, reality for the technologists was something that could be manipulated and molded into anything they wanted.

THEATER OF THE ABSURD

One technologist in particular rose to power. And his plans and schemes became crucial to Putin's control of Russia. His name was Vladislav Surkov. Surkov originally worked in theater, and academics who've analyzed his path to power say that Surkov took avant-garde theater ideas and transplanted them into the beating heart of Russian politics. The goal wasn't just to manipulate political audiences, but to drill down further and toy with their very idea of the world so they would become unsure of reality. In short, Surkov transformed Russian politics into a puzzling and forever morphing piece of theater.

Surkov used Kremlin cash to fund all sorts of political factions. These ranged from mass anti-fascist youth organizations to the total opposite in neo-Nazi skinheads. Surkov even sponsored entire political parties which were opposed to Putin. And then, get this, as if he was announcing the fact that folks were about to be assimilated, Surkov told them what he was doing. That's right, he announced his political trickery which meant nobody was really sure what was real or what was fake in modern Russia. Surkov's was a strategy of power that kept any opposition constantly

confused. It was politics as a ceaseless shape-shifting entity, as unstoppable as Species 8472 because it was indefinable, and as assimilating as the Borg. Meanwhile, real power was hidden away behind the stage, wielded without anyone witnessing it.

THE DAYS HAVE GONE DOWN IN THE WEST

And then the same thing started happening in the West. Here too it was becoming obvious that the system had fatal flaws. It seemed that almost every week there were new scandals, such as banks' collusion in global corruption, or colossal tax avoidance by most of the major corporations, or the clandestine Borg-like surveillance of everyone's emails by the National Security Agency. And yet, not one truly responsible soul was prosecuted, save for a few minions at the lowest levels. And despite all this, behind the scenes, the huge inequalities kept growing, the power structure stayed the same. Nothing changed, as change would destabilize the system.

MAKE AMERICA GREAT AGAIN

To help sew confusion and a sense of unreality about politics, rhetoric is creatively used. Slogans are at times deliberately ambiguous or vague. Consider, for example, the Delphic campaign slogan, "Make America Great Again." The slogan doesn't say how this will be done. Nor does it detail what will be great in this nebulous and fuzzy future that isn't so great now. And so the sentiment becomes difficult to disagree with. People are left to think that whatever it is they believe needs to be made "great again" will be so. They are assimilated.

People became assimilated into the Trump collective. He and his audience were aware that much of what he said had a very shaky relationship with reality. And this had an effect on science and journalism. That's because the core belief of the scientist and the journalist is to assert the truth and expose falsehoods. Journalism, like science, is evidentiary. But, with Trump, evidence became irrelevant. Liberals were outraged. And yet they voiced their anger not in real life, as it were, but in that other great sci-fi invention, cyberspace. And so their anger had little or no effect—the corporate algorithms ensured that Trump's opponents spoke only to those who already agreed with them. Rather, and ironically, the waves of protest benefitted the large corporations who ran the social media platforms. As one analyst cynically said, "Angry people click more." And so the radical fury that may have changed the world in the past was now simply a fuel that was feeding new systems of power.

ON BEING TRULY EDUCATED

Why does this matter? And why does it mean many humans are becoming Borg-like? Well, let's think about what it means for an individual in a democratic society to be truly educated. The kind of society that features in the *Star Trek* future. To answer the question about what it means to be truly educated we could go back to the views expressed by Prussian philosopher Wilhelm von Humboldt. Von Humboldt was the architect of the Prussian public education model, a system also used in the United States and Japan. He was a leading humanist figure of the Enlightenment who wrote extensively on education and human development. And he argued that the core principle and requirement of a fulfilled human being is the ability to inquire and create constructively, independently without external controls. Exactly those kinds of external controls parodied by Orwell, used in modern politics, and characterized by the Borg.

A modern scientist would say much the same as von Humboldt, for within modern courses on science, it's not so much what's covered in the course that's crucial, it's more important what you actually discover. To be truly educated in science is to be in a position to inquire and create on the basis of the resources available to you. Resources that you know are reliable and you can refer to when you need to formulate solutions to serious questions. But what if those resources are compromised? What if the independence of the facts are called into question? To independently deal with the challenges that the world presents, and to develop in the course of your self-education and inquiry, in cooperation and solidarity with others, that's what an educational system should cultivate from kindergarten to college, especially in science. As Albert Einstein put it, "How far superior an education that stresses independent action and personal responsibility is to one that relies on drill, external authority, and ambition."

And that's what's under threat around the world today. The age of reason seems to be ending. Resistance seems futile, as hope is at a premium. Wisdom and knowledge are called into question as science is downgraded or dissed. Democracy, based on open truths, is drowning, while autocracy, based on lies and disinformation, is on the march.

STAR TREK'S GOSPEL FOR OUTER SPACE

We spoke at the very beginning of this book about *Star Trek*'s most enduring legacies, especially diversity and inclusivity. With modern politics and the Borg in mind, let's look again at what Gene Roddenberry considered the *Star Trek* philosophy to be in 1976: "*Star Trek* is a cult. Television has incredible power . . . What's to keep selfish interests from creating other cults, for selfish purposes, industrial cartels, political parties, and governments? Ultimate power in this world, as you know, has always been one simple thing: the control and manipulation of minds." What a wonderful contrast there is between the inclusivity and multiculturalism of *Star Trek* and the monolithic uniculturalism of the Borg. This is what Roddenberry had to say about technological control getting into the wrong hands: "The worst possible thing that can happen to all of us is for the future to somehow press us into a common mold where we begin to act and talk and look and think alike."

Finally, let's leave the last few words of this book to Gene Roddenberry in the hope of a brighter future: "We believed that the often-ridiculed mass audience is sick of this world's petty nationalism and all its old ways and old hatreds, and that people are not only willing but anxious to think beyond most petty beliefs that have for so long kept mankind divided. *Star Trek* was an attempt to say that humanity will reach maturity and wisdom on the day that it begins not just to tolerate, but take a special delight in differences in ideas and differences in life forms . . . What Star Trek proves . . . is that the much-maligned common man and common woman has an enormous hunger for brotherhood. They are ready for the twenty-third century now, and they are light-years ahead of their petty governments and their visionless leaders."

INDEX